# Kawasaki AE/AR50 & 80 Owners Workshop Manual

## by Chris Rogers
with an additional Chapter on the 1984 on models
## by Jeremy Churchill

**Models covered**
AE50 A. 49cc. UK April 1981 to July 1985
AR50 A. 49cc. UK March 1981 to January 1983, US October 1981 to 1982
AR50 C. 49cc. UK January 1983 on
AE80 A. 78cc. UK March 1981 to January 1983
AE80 B. 78cc. UK January 1983 to March 1987
AR80 A. 78cc. UK April 1981 to January 1983, US October 1981 to 1982
AR80 C. 78cc. UK January 1983 to November 1992

**ISBN 978 1 85960 173 0**

Printed in the UK    (1007-7U7)

ABCDE
FGHIJ
KLMNO
PQRS
2

**Haynes Publishing Group**
Sparkford Nr Yeovil
Somerset BA22 7JJ England

**Haynes Publications, Inc**
859 Lawrence Drive
Newbury Park
California 91320 USA

| British Library Cataloguing in Publication Data |
| --- |
| A Catalogue record for this book is available from the British Library |

| Library of Congress |
| --- |
| 95-079303 |

# Acknowledgements

Special thanks are due to the Taunton Kawasaki Centre for supplying the AE50 and AR80 machines featured in Chapters 1 to 6 of this manual, and to CW Motorcycles of Dorchester, who supplied the AR50 C6 featured on the front cover.

Kawasaki Motors (UK) Limited gave permission to use the line drawings contained in this Manual and supplied much of the technical information used by the Author.

Thanks are also due to the Avon Rubber Company, who kindly supplied information and technical assistance on tyre fitting; NGK Spark Plugs (UK) Ltd for information on spark plug maintenance and electrode conditions, and Renold Ltd for advice on chain care and renewal.

# About this manual

The purpose of this manual is to present the owner with a concise and graphic guide which will enable him to tackle any operation from basic routine maintenance to a major overhaul. It has been assumed that any work would be undertaken without the luxury of a well-equipped workshop and a range of manufacturer's service tools.

To this end, the machine featured in the manual was stripped and rebuilt in our own workshop, by a team comprising a mechanic, a photographer and the author. The resulting photographic sequence depicts events as they took place, the hands shown being those of the author and the mechanic.

The use of specialised, and expensive, service tools was avoided unless their use was considered to be essential due to risk of breakage or injury. There is usually some way of improvising a method of removing a stubborn component, providing that a suitable degree of care is exercised.

The author learnt his motorcycle mechanics over a number of years, faced with the same difficulties and using similar facilities to those encountered by most owners. It is hoped that this practical experience can be passed on through the pages of this manual.

Where possible, a well-used example of the machine is chosen for the workshop project, as this highlights any areas which might be particularly prone to giving rise to problems. In this way, any such difficulties are encountered and resolved before the text is written, and the techniques used to deal with them can be incorporated in the relevant section. Armed with a working knowledge of the machine, the author undertakes a considerable amount of research in order that the maximum amount of data can be included in the manual.

A comprehensive section, preceding the main part of the manual, describes procedures for carrying out the routine maintenance of the machine at intervals of time and mileage. This section is included particularly for those owners who wish to ensure the efficient day-to-day running of their motorcycle, but who choose not to undertake overhaul or renovation work.

Each Chapter is divided into numbered sections. Within these sections are numbered paragraphs. Cross reference throughout the manual is quite straightforward and logical. When reference is made 'See Section 6.10' it means Section 6, paragraph 10 in the same Chapter. If another Chapter were intended, the reference would read, for example, 'See Chapter 2, Section 6.10'. All the photographs are captioned with a section/paragraph number to which they refer and are relevant to the Chapter text adjacent.

Figures (usually line illustrations) appear in a logical but numerical order, within a given Chapter. Fig. 1.1 therefore refers to the first figure in Chapter 1.

Left-hand and right-hand descriptions of the machines and their components refer to the left and right of a given machine when the rider is seated normally.

Motorcycle manufacturers continually make changes to specifications and recommendations, and these, when notified, are incorporated into our manuals at the earliest opportunity.

We take great pride in the accuracy of information given in this manual, but motorcycle manufacturers make alterations and design changes during the production run of a particular motorcycle of which they do not inform us. No liability can be accepted by the authors or publishers for loss, damage or injury caused by any errors in, or omissions from, the information given.

# Contents

The Kawasaki AE50 A2 model

The Kawasaki AE80 B1 model

The Kawasaki AR50 A1 model

The Kawasaki AR80 C1 model

# Introduction to the Kawasaki AE/AR 50 & 80 models

The machines covered in this manual all share the same basic engine/gearbox unit (apart from the obvious capacity difference and the fact that the UK 50 models use a 5-speed transmission and an engine with a restricted output to comply with legal requirements) and very similar cycle parts. The AE trail models can be identified by their 'off-road' styling, drum brakes and wire-spoked wheels, while the AR road models feature a front disc brake, cast alloy wheels and racetrack-inspired styling, including a small front fork-mounted fairing.

Although these machines have remained substantially unchanged, various minor modifications were made which require either an altered working procedure or revised specifications; making it essential for the owner to identify his or her machine exactly. Kawasaki identify their models first by a suffix letter, then by a suffix number; this number is usually related to a production year. Given below are the frame and engine numbers (where available) at which production of each model commenced, also the approximate dates of import and any significant identifying features. Note that a machine's date of sale need not necessarily coincide with the date of import given below.

**AE50 A1** – Production year 1981/2, imported into the UK from April 1981 to January 1983. Frame number AE050A-000001 on. Engine number AR050AE000001 on. Available in Firecracker Red, Lime Green or Brilliant Blue.

**AE50 A2** – Production year 1983, imported into the UK from January 1983 to July 1985. Frame number AE050A-007401 on. Engine number AR050AE023831 on. Available in Sunbeam Red, Lime Green or Polar White. Identical to A1 apart from changes to graphics on fuel tank, side panels and seat, also modified wheel spindles, brake components and control cables.

**AR50 A1** – Production year 1981/2, imported into the UK from March 1981 to January 1983. Frame number AR050A-000001 on. Engine number AR050AE000001 on. Available in Firecracker Red, Lime Green or Polar White.

**AR50 A1** – Production year 1982, imported into the US from October 1981 to 1982. Frame numbers JKAARYA1*CA013013 to 014042. Engine numbers AR050AE019830 to 021001. Available in Firecracker Red.

**AR50 C1** – Production year 1983, imported into the UK from January to November 1983. Frame number AR050C-000001 on. Engine number AR050AE027891 on. Available in Sunbeam Red or Ebony. Similar to A1 but features GPz-style side panels, seat and tail unit.

**AE80 A1** – Production year 1981/2, imported into the UK from March 1981 to January 1983. Frame number AE080A-000001 on. Engine number AR080AE000001 on. Available in Firecracker Red or Brilliant Blue.

**AE80 B1** – Production year 1983, imported into the UK from January 1983 to March 1987. Frame number AE080B-000001 on. Engine number AR080AE000001 on. Available in Sunbeam Red, Lime Green or Polar White. Identical to A1 apart from change to graphics on fuel tank, side panels and seat.

**AR80 A1** – Production year 1981/2, imported into the UK from April 1981 to January 1983. Frame number AR080A-000001 on. Engine number AR080AE000001 on. Available in Firecracker Red, Lime Green or Polar White.

**AR80 A1** – Production year 1982, imported into the US (except California) from October 1981 to 1982. Frame numbers JKAARWA1*CA010301 to 014283. Engine numbers AR080AE017108 to 023361. Available in Firecracker Red.

**AR80 C1** – Production year 1983, imported into the UK from January 1983 to January 1984. Frame number AR080C-000001 on. Engine number AR080AE026861 on. Available in Sunbeam Red or Polar White. Similar to A1 but features GPz-style side panels, seat and tail unit.

*Note: The digit indicated by the asterisk in the (frame) VIN numbers of the US models will vary from machine to machine.*

*For models produced from 1984 on, namely the UK AR50 and 80 C2, C3, C4, C5, C6, C7, C8, C9 and C10, refer to Chapter 7.*

# Model dimensions and weights

| | AR models | AE models |
| --- | --- | --- |
| Overall length | 1855 mm (73.0 in) – A1 models<br>1880 mm (74.0 in) – C1 models | 1880 mm (74.0 in) |
| Overall width | 630 mm (24.8 in) | 785 mm (30.9 in) |
| Overall height | 1145 mm (45.1 in) | 1050 mm (41.3 in) |
| Wheelbase:<br>50<br>80 | 1195 mm (47.0 in)<br>1205 mm (47.4 in) – UK models<br>1200 mm (47.2 in) – US models | 1195 mm (47.0 in)<br>1205 mm (47.4 in) |
| Ground clearance | 175 mm (6.9 in) | 240 mm (9.4 in) |
| Seat height | 790 mm (31.1 in) – A1 models<br>785 mm (30.9 in) – C1 models | 795 mm (31.3 in) |
| Dry weight | 75 kg (165 lb) – A1 models<br>78 kg (172 lb) – C1 models | 77 kg (169.7 lb) |

# Ordering spare parts

When ordering spare parts for any Kawasaki, deal direct with an official Kawasaki agent who should be able to supply most of the parts ex-stock. Parts cannot be obtained from Kawasaki direct, even if the parts required are not held in stock. Always quote the engine and frame numbers in full, especially if parts are required for earlier models.

The frame number is stamped on the steering head and the engine number on the top surface of the right-hand crankcase half.

Use only genuine Kawasaki spares. Some pattern parts are available that may be packed in similar looking packages. They should only be used if genuine parts are hard to obtain or in an emergency, for they do not normally last as long as genuine parts, even although there may be a price advantage.

Some of the more expendable parts such as spark plugs, bulbs, oils and greases etc, can be obtained from accessory shops and motor factors, who have convenient opening hours, and can be found not far from home. It is also possible to obtain parts on a Mail Order basis from a number of specialists who advertise regularly in the motorcycle magazines.

Location of frame number

Location of engine number

# Safety first!

Professional motor mechanics are trained in safe working procedures. However enthusiastic you may be about getting on with the job in hand, do take the time to ensure that your safety is not put at risk. A moment's lack of attention can result in an accident, as can failure to observe certain elementary precautions.

There will always be new ways of having accidents, and the following points do not pretend to be a comprehensive list of all dangers; they are intended rather to make you aware of the risks and to encourage a safety-conscious approach to all work you carry out on your vehicle.

### Essential DOs and DON'Ts

**DON'T** start the engine without first ascertaining that the transmission is in neutral.

**DON'T** suddenly remove the filler cap from a hot cooling system – cover it with a cloth and release the pressure gradually first, or you may get scalded by escaping coolant.

**DON'T** attempt to drain oil until you are sure it has cooled sufficiently to avoid scalding you.

**DON'T** grasp any part of the engine, exhaust or silencer without first ascertaining that it is sufficiently cool to avoid burning you.

**DON'T** allow brake fluid or antifreeze to contact the machine's paintwork or plastic components.

**DON'T** syphon toxic liquids such as fuel, brake fluid or antifreeze by mouth, or allow them to remain on your skin.

**DON'T** inhale dust – it may be injurious to health (see *Asbestos* heading).

**DON'T** allow any spilt oil or grease to remain on the floor – wipe it up straight away, before someone slips on it.

**DON'T** use ill-fitting spanners or other tools which may slip and cause injury.

**DON'T** attempt to lift a heavy component which may be beyond your capability – get assistance.

**DON'T** rush to finish a job, or take unverified short cuts.

**DON'T** allow children or animals in or around an unattended vehicle.

**DON'T** inflate a tyre to a pressure above the recommended maximum. Apart from overstressing the carcase and wheel rim, in extreme cases the tyre may blow off forcibly.

**DO** ensure that the machine is supported securely at all times. This is especially important when the machine is blocked up to aid wheel or fork removal.

**DO** take care when attempting to slacken a stubborn nut or bolt. It is generally better to pull on a spanner, rather than push, so that if slippage occurs you fall away from the machine rather than on to it.

**DO** wear eye protection when using power tools such as drill, sander, bench grinder etc.

**DO** use a barrier cream on your hands prior to undertaking dirty jobs – it will protect your skin from infection as well as making the dirt easier to remove afterwards; but make sure your hands aren't left slippery. Note that long-term contact with used engine oil can be a health hazard.

**DO** keep loose clothing (cuffs, tie etc) and long hair well out of the way of moving mechanical parts.

**DO** remove rings, wristwatch etc, before working on the vehicle – especially the electrical system.

**DO** keep your work area tidy – it is only too easy to fall over articles left lying around.

**DO** exercise caution when compressing springs for removal or installation. Ensure that the tension is applied and released in a controlled manner, using suitable tools which preclude the possibility of the spring escaping violently.

**DO** ensure that any lifting tackle used has a safe working load rating adequate for the job.

**DO** get someone to check periodically that all is well, when working alone on the vehicle.

**DO** carry out work in a logical sequence and check that everything is correctly assembled and tightened afterwards.

**DO** remember that your vehicle's safety affects that of yourself and others. If in doubt on any point, get specialist advice.

**IF,** in spite of following these precautions, you are unfortunate enough to injure yourself, seek medical attention as soon as possible.

### Asbestos

Certain friction, insulating, sealing, and other products – such as brake linings, clutch linings, gaskets, etc – contain asbestos. *Extreme care must be taken to avoid inhalation of dust from such products since it is hazardous to health.* If in doubt, assume that they *do* contain asbestos.

### Fire

Remember at all times that petrol (gasoline) is highly flammable. Never smoke, or have any kind of naked flame around, when working on the vehicle. But the risk does not end there – a spark caused by an electrical short-circuit, by two metal surfaces contacting each other, by careless use of tools, or even by static electricity built up in your body under certain conditions, can ignite petrol vapour, which in a confined space is highly explosive.

Always disconnect the battery earth (ground) terminal before working on any part of the fuel or electrical system, and never risk spilling fuel on to a hot engine or exhaust.

It is recommended that a fire extinguisher of a type suitable for fuel and electrical fires is kept handy in the garage or workplace at all times. Never try to extinguish a fuel or electrical fire with water.

**Note:** *Any reference to a 'torch' appearing in this manual should always be taken to mean a hand-held battery-operated electric lamp or flashlight. It does **not** mean a welding/gas torch or blowlamp.*

### Fumes

Certain fumes are highly toxic and can quickly cause unconsciousness and even death if inhaled to any extent. Petrol (gasoline) vapour comes into this category, as do the vapours from certain solvents such as trichloroethylene. Any draining or pouring of such volatile fluids should be done in a well ventilated area.

When using cleaning fluids and solvents, read the instructions carefully. Never use materials from unmarked containers – they may give off poisonous vapours.

Never run the engine of a motor vehicle in an enclosed space such as a garage. Exhaust fumes contain carbon monoxide which is extremely poisonous; if you need to run the engine, always do so in the open air or at least have the rear of the vehicle outside the workplace.

### The battery

Never cause a spark, or allow a naked light, near the vehicle's battery. It will normally be giving off a certain amount of hydrogen gas, which is highly explosive.

Always disconnect the battery earth (ground) terminal before working on the fuel or electrical systems.

If possible, loosen the filler plugs or cover when charging the battery from an external source. Do not charge at an excessive rate or the battery may burst.

Take care when topping up and when carrying the battery. The acid electrolyte, even when diluted, is very corrosive and should not be allowed to contact the eyes or skin.

If you ever need to prepare electrolyte yourself, always add the acid slowly to the water, and never the other way round. Protect against splashes by wearing rubber gloves and goggles.

### Mains electricity and electrical equipment

When using an electric power tool, inspection light etc, always ensure that the appliance is correctly connected to its plug and that, where necessary, it is properly earthed (grounded). Do not use such appliances in damp conditions and, again, beware of creating a spark or applying excessive heat in the vicinity of fuel or fuel vapour. Also ensure that the appliances meet the relevant national safety standards.

### Ignition HT voltage

A severe electric shock can result from touching certain parts of the ignition system, such as the HT leads, when the engine is running or being cranked, particularly if components are damp or the insulation is defective. Where an electronic ignition system is fitted, the HT voltage is much higher and could prove fatal.

# Tools and working facilities

The first priority when undertaking maintenance or repair work of any sort on a motorcycle is to have a clean, dry, well-lit working area. Work carried out in peace and quiet in the well-ordered atmosphere of a good workshop will give more satisfaction and much better results than can usually be achieved in poor working conditions. A good workshop must have a clean flat workbench or a solidly constructed table of convenient working height. The workbench or table should be equipped with a vice which has a jaw opening of at least 4 in (100 mm). A set of jaw covers should be made from soft metal such as aluminium alloy or copper, or from wood. These covers will minimise the marking or damaging of soft or delicate components which may be clamped in the vice. Some clean, dry, storage space will be required for tools, lubricants and dismantled components. It will be necessary during a major overhaul to lay out engine/gearbox components for examination and to keep them where they will remain undisturbed for as long as is necessary. To this end it is recommended that a supply of metal or plastic containers of suitable size is collected. A supply of clean, lint-free, rags for cleaning purposes and some newspapers, other rags, or paper towels for mopping up spillages should also be kept. If working on a hard concrete floor note that both the floor and one's knees can be protected from oil spillages and wear by cutting open a large cardboard box and spreading it flat on the floor under the machine or workbench. This also helps to provide some warmth in winter and to prevent the loss of nuts, washers, and other tiny components which have a tendency to disappear when dropped on anything other than a perfectly clean, flat, surface.

Unfortunately, such working conditions are not always available to the home mechanic. When working in poor conditions it is essential to take extra time and care to ensure that the components being worked on are kept scrupulously clean and to ensure that no components or tools are lost or damaged.

A selection of good tools is a fundamental requirement for anyone contemplating the maintenance and repair of a motor vehicle. For the owner who does not possess any, their purchase will prove a considerable expense, offsetting some of the savings made by doing-it-yourself. However, provided that the tools purchased meet the relevant national safety standards and are of good quality, they will last for many years and prove an extremely worthwhile investment.

To help the average owner to decide which tools are needed to carry out the various tasks detailed in this manual, we have compiled three lists of tools under the following headings: *Maintenance and minor repair, Repair and overhaul,* and *Specialized.* The newcomer to practical mechanics should start off with the simpler jobs around the vehicle. Then, as his confidence and experience grow, he can undertake more difficult tasks, buying extra tools as and when they are needed. In this way, a *Maintenance and minor repair* tool kit can be built-up into a *Repair and overhaul* tool kit over a considerable period of time without any major cash outlays. The experienced home mechanic will have a tool kit good enough for most repair and overhaul procedures and will add tools from the specialized category when he feels the expense is justified by the amount of use these tools will be put to.

It is obviously not possible to cover the subject of tools fully here. For those who wish to learn more about tools and their use there is a book entitled *Motorcycle Workshop Practice Manual* available from the publishers of this manual.

As a general rule, it is better to buy the more expensive, good quality tools. Given reasonable use, such tools will last for a very long time, whereas the cheaper, poor quality, item will wear out faster and need to be renewed more often, thus nullifying the original saving. There is also the risk of a poor quality tool breaking while in use, causing personal injury or expensive damage to the component being worked on.

For practically all tools, a tool factor is the best source since he will have a very comprehensive range compared with the average garage or accessory shop. Having said that, accessory shops often offer excellent quality tools at discount prices, so it pays to shop around. There are plenty of tools around at reasonable prices, but always aim to purchase items which meet the relevant national safety standards. If in doubt, seek the advice of the shop proprietor or manager before making a purchase.

The basis of any toolkit is a set of spanners. While open-ended spanners with their slim jaws, are useful for working on awkwardly-positioned nuts, ring spanners have advantages in that they grip the nut far more positively. There is less risk of the spanner slipping off the nut and damaging it, for this reason alone ring spanners are to be preferred. Ideally, the home mechanic should acquire a set of each, but if expense rules this out a set of combination spanners (open-ended at one end and with a ring of the same size at the other) will provide a good compromise. Another item which is so useful it should be considered an essential requirement for any home mechanic is a set of socket spanners. These are available in a variety of drive sizes. It is recommended that the ½-inch drive type is purchased to begin with as although bulkier and more expensive than the ⅜-inch type, the larger size is far more common and will accept a greater variety of torque wrenches, extension pieces and socket sizes. The socket set should comprise sockets of sizes between 8 and 24 mm, a reversible ratchet drive, an extension bar of about 10 inches in length, a spark plug socket with a rubber insert, and a universal joint. Other attachments can be added to the set at a later date.

## Maintenance and minor repair tool kit

Set of spanners 8 – 24 mm
Set of sockets and attachments
Spark plug spanner with rubber insert – 10, 12, or 14 mm as appropriate
Adjustable spanner
C-spanner/pin spanner
Torque wrench (same size drive as sockets)
Set of screwdrivers (flat blade)
Set of screwdrivers (cross-head)
Set of Allen keys 4 – 10 mm
Impact screwdriver and bits
Ball pein hammer – 2 lb
Hacksaw (junior)
Self-locking pliers – Mole grips or vice grips
Pliers – combination
Pliers – needle nose
Wire brush (small)
Soft-bristled brush
Tyre pump
Tyre pressure gauge
Tyre tread depth gauge
Oil can
Fine emery cloth
Funnel (medium size)
Drip tray
Grease gun
Set of feeler gauges
Brake bleeding kit
Strobe timing light
Continuity tester (dry battery and bulb)
Soldering iron and solder
Wire stripper or craft knife
PVC insulating tape
Assortment of split pins, nuts, bolts, and washers

## Repair and overhaul toolkit

The tools in this list are virtually essential for anyone undertaking major repairs to a motorcycle and are additional to the tools listed above.

Plastic or rubber soft-faced mallet
Pliers – electrician's side cutters
Circlip pliers – internal (straight or right-angled tips are available)
Circlip pliers – external
Cold chisel
Centre punch
Pin punch
Scriber
Scraper (made from soft metal such as aluminium or copper)
Soft metal drift
Steel rule/straight edge
Assortment of files

Electric drill and bits
Wire brush (large)
Soft wire brush (similar to those used for cleaning suede shoes)
Sheet of plate glass
Hacksaw (large)
Stud extractor set (E-Z out)

## Specialized tools

This is not a list of the tools made by the machine's manufacturer to carry out a specific task on a limited range of models. Occasional references are made to such tools in the text of this manual and, in general, an alternative method of carrying out the task without the manufacturer's tool is given where possible. The tools mentioned in this list are those which are not used regularly and are expensive to buy in view of their infrequent use. Where this is the case it may be possible to hire or borrow the tools against a deposit from a local dealer or tool hire shop. An alternative is for a group of friends or a motorcycle club to join in the purchase.

Piston ring compressor
Universal bearing puller
Cylinder bore honing attachment (for electric drill)
Micrometer set
Vernier calipers
Dial gauge set
Cylinder compression gauge
Multimeter
Dwell meter/tachometer

## Care and maintenance of tools

Whatever the quality of the tools purchased, they will last much longer if cared for. This means in practice ensuring that a tool is used for its intended purpose; for example screwdrivers should not be used as a substitute for a centre punch, or as chisels. Always remove dirt or grease and any metal particles but remember that a light film of oil will prevent rusting if the tools are infrequently used. The common tools can be kept together in a large box or tray but the more delicate, and more expensive, items should be stored separately where they cannot be damaged. When a tool is damaged or worn out, be sure to renew it immediately. It is false economy to continue to use a worn spanner or screwdriver which may slip and cause expensive damage to the component being worked on.

## Fastening systems

Fasteners, basically, are nuts, bolts and screws used to hold two or more parts together. There are a few things to keep in mind when working with fasteners. Almost all of them use a locking device of some type; either a lock washer, lock nut, locking tab or thread adhesive. All threaded fasteners should be clean, straight, have undamaged threads and undamaged corners on the hexagon head where the spanner fits. Develop the habit of replacing all damaged nuts and bolts with new ones.

Rusted nuts and bolts should be treated with a rust penetrating fluid to ease removal and prevent breakage. After applying the rust penetrant, let it 'work' for a few minutes before trying to loosen the nut or bolt. Badly rusted fasteners may have to be chiseled off or removed with a special nut breaker, available at tool shops.

Flat washers and lock washers, when removed from an assembly should always be replaced exactly as removed. Replace any damaged washers with new ones. Always use a flat washer between a lock washer and any soft metal surface (such as aluminium), thin sheet metal or plastic. Special lock nuts can only be used once or twice before they lose their locking ability and must be renewed.

If a bolt or stud breaks off in an assembly, it can be drilled out and removed with a special tool called an E-Z out. Most dealer service departments and motorcycle repair shops can perform this task, as well as others (such as the repair of threaded holes that have been stripped out).

# Spanner size comparison

| Jaw gap (in) | Spanner size |
|---|---|
| 0.250 | $\frac{1}{4}$ in AF |
| 0.276 | 7 mm |
| 0.313 | $\frac{5}{16}$ in AF |
| 0.315 | 8 mm |
| 0.344 | $\frac{11}{32}$ in AF; $\frac{1}{8}$ in Whitworth |
| 0.354 | 9 mm |
| 0.375 | $\frac{3}{8}$ in AF |
| 0.394 | 10 mm |
| 0.433 | 11 mm |
| 0.438 | $\frac{7}{16}$ in AF |
| 0.445 | $\frac{3}{16}$ in Whitworth; $\frac{1}{4}$ in BSF |
| 0.472 | 12 mm |
| 0.500 | $\frac{1}{2}$ in AF |
| 0.512 | 13 mm |
| 0.525 | $\frac{1}{4}$ In Whitworth; $\frac{5}{16}$ in BSF |
| 0.551 | 14 mm |
| 0.563 | $\frac{9}{16}$ in AF |
| 0.591 | 15 mm |
| 0.600 | $\frac{5}{16}$ in Whitworth; $\frac{3}{8}$ in BSF |
| 0.625 | $\frac{5}{8}$ in AF |
| 0.630 | 16 mm |
| 0.669 | 17 mm |
| 0.686 | $\frac{11}{16}$ in AF |
| 0.709 | 18 mm |
| 0.710 | $\frac{3}{8}$ in Whitworth; $\frac{7}{16}$ in BSF |
| 0.748 | 19 mm |
| 0.750 | $\frac{3}{4}$ in AF |
| 0.813 | $\frac{13}{16}$ in AF |
| 0.820 | $\frac{7}{16}$ in Whitworth; $\frac{1}{2}$ in BSF |
| 0.866 | 22 mm |
| 0.875 | $\frac{7}{8}$ in AF |
| 0.920 | $\frac{1}{2}$ in Whitworth; $\frac{9}{16}$ in BSF |
| 0.938 | $\frac{15}{16}$ in AF |
| 0.945 | 24 mm |
| 1.000 | 1 in AF |
| 1.010 | $\frac{9}{16}$ in Whitworth; $\frac{5}{8}$ in BSF |
| 1.024 | 26 mm |
| 1.063 | $1\frac{1}{16}$ in AF; 27 mm |
| 1.100 | $\frac{5}{8}$ in Whitworth; $\frac{11}{16}$ in BSF |
| 1.125 | $1\frac{1}{8}$ in AF |
| 1.181 | 30 mm |
| 1.200 | $\frac{11}{16}$ in Whitworth; $\frac{3}{4}$ in BSF |
| 1.250 | $1\frac{1}{4}$ in AF |
| 1.260 | 32 mm |
| 1.300 | $\frac{3}{4}$ in Whitworth; $\frac{7}{8}$ in BSF |
| 1.313 | $1\frac{5}{16}$ in AF |
| 1.390 | $\frac{13}{16}$ in Whitworth; $\frac{15}{16}$ in BSF |
| 1.417 | 36 mm |
| 1.438 | $1\frac{7}{16}$ in AF |
| 1.480 | $\frac{7}{8}$ in Whitworth; 1 in BSF |
| 1.500 | $1\frac{1}{2}$ in AF |
| 1.575 | 40 mm; $\frac{15}{16}$ in Whitworth |
| 1.614 | 41 mm |
| 1.625 | $1\frac{5}{8}$ in AF |
| 1.670 | 1 in Whitworth; $1\frac{1}{8}$ in BSF |
| 1.688 | $1\frac{11}{16}$ in AF |
| 1.811 | 46 mm |
| 1.813 | $1\frac{13}{16}$ in AF |
| 1.860 | $1\frac{1}{8}$ in Whitworth; $1\frac{1}{4}$ in BSF |
| 1.875 | $1\frac{7}{8}$ in AF |
| 1.969 | 50 mm |
| 2.000 | 2 in AF |
| 2.050 | $1\frac{1}{4}$ in Whitworth; $1\frac{3}{8}$ in BSF |
| 2.165 | 55 mm |
| 2.362 | 60 mm |

# Standard torque settings

Specific torque settings will be found at the end of the specifications section of each chapter. Where no figure is given, bolts should be secured according to the table below.

| Fastener type (thread diameter) | kgf m | lbf ft |
|---|---|---|
| 5mm bolt or nut | 0.45 – 0.6 | 3.5 – 4.5 |
| 6 mm bolt or nut | 0.8 – 1.2 | 6 – 9 |
| 8 mm bolt or nut | 1.8 – 2.5 | 13 – 18 |
| 10 mm bolt or nut | 3.0 – 4.0 | 22 – 29 |
| 12 mm bolt or nut | 5.0 – 6.0 | 36 – 43 |
| 5 mm screw | 0.35 – 0.5 | 2.5 – 3.6 |
| 6 mm screw | 0.7 – 1.1 | 5 – 8 |
| 6 mm flange bolt | 1.0 – 1.4 | 7 – 10 |
| 8 mm flange bolt | 2.4 – 3.0 | 17 – 22 |
| 10 mm flange bolt | 3.0 – 4.0 | 22 – 29 |

# Choosing and fitting accessories

The range of accessories available to the modern motorcyclist is almost as varied and bewildering as the range of motorcycles. This Section is intended to help the owner in choosing the correct equipment for his needs and to avoid some of the mistakes made by many riders when adding accessories to their machines. It will be evident that the Section can only cover the subject in the most general terms and so it is recommended that the owner, having decided that he wants to fit, for example, a luggage rack or carrier, seeks the advice of several local dealers and the owners of similar machines. This will give a good idea of what makes of carrier are easily available, and at what price. Talking to other owners will give some insight into the drawbacks or good points of any one make. A walk round the motorcycles in car parks or outside a dealer will often reveal the same sort of information.

The first priority when choosing accessories is to assess exactly what one needs. It is, for example, pointless to buy a large heavy-duty carrier which is designed to take the weight of fully laden panniers and topbox when all you need is a place to strap on a set of waterproofs and a lunchbox when going to work. Many accessory manufacturers have ranges of equipment to cater for the individual needs of different riders and this point should be borne in mind when looking through a dealer's catalogues. Having decided exactly what is required and the use to which the accessories are going to be put, the owner will need a few hints on what to look for when making the final choice. To this end the Section is now sub-divided to cover the more popular accessories fitted. Note that it is in no way a customizing guide, but merely seeks to outline the practical considerations to be taken into account when adding aftermarket equipment to a motorcycle.

## Fairings and windscreens

A fairing is possibly the single, most expensive, aftermarket item to be fitted to any motorcycle and, therefore, requires the most thought before purchase. Fairings can be divided into two main groups: front fork mounted handlebar fairings and windscreens, and frame mounted fairings.

The first group, the front fork mounted fairings, are becoming far more popular than was once the case, as they offer several advantages over the second group. Front fork mounted fairings generally are much easier and quicker to fit, involve less modification to the motorcycle, do not as a rule restrict the steering lock, permit a wider selection of handlebar styles to be used, and offer adequate protection for much less money than the frame mounted type. They are also lighter, can be swapped easily between different motorcycles, and are available in a much greater variety of styles. Their main disadvantages are that they do not offer as much weather protection as the frame mounted types, rarely offer any storage space, and, if poorly fitted or naturally incompatible, can have an adverse effect on the stability of the motorcycle.

The second group, the frame mounted fairings, are secured so rigidly to the main frame of the motorcycle that they can offer a substantial amount of protection to motorcycle and rider in the event of a crash. They offer almost complete protection from the weather and, if double-skinned in construction, can provide a great deal of useful storage space. The feeling of peace, quiet and complete relaxation encountered when riding behind a good full fairing has to be experienced to be believed. For this reason full fairings are considered essential by most touring motorcyclists and by many people who ride all year round. The main disadvantages of this type are that fitting can take a long time, often involving removal or modification of standard motorcycle components, they restrict the steering lock and they can add up to about 40 lb to the weight of the machine. They do not usually affect the stability of the machine to any great extent once the front tyre pressure and suspension have been adjusted to compensate for the extra weight, but can be affected by sidewinds.

The first thing to look for when purchasing a fairing is the quality of the fittings. A good fairing will have strong, substantial brackets constructed from heavy-gauge tubing; the brackets must be shaped to fit the frame or forks evenly so that the minimum of stress is imposed on the assembly when it is bolted down. The brackets should be properly painted or finished – a nylon coating being the favourite of the better manufacturers – the nuts and bolts provided should be of the same thread and size standard as is used on the motorcycle and be properly plated. Look also for shakeproof locking nuts or locking washers to ensure that everything remains securely tightened down. The fairing shell is generally made from one of two materials: fibreglass or ABS plastic. Both have their advantages and disadvantages, but the main consideration for the owner is that fibreglass is much easier to repair in the event of damage occurring to the fairing. Whichever material is used, check that it is properly finished inside as well as out, that the edges are protected by beading and that the fairing shell is insulated from vibration by the use of rubber grommets at all mounting points. Also be careful to check that the windscreen is retained by plastic bolts which will snap on impact so that the windscreen will break away and not cause personal injury in the event of an accident.

Having purchased your fairing or windscreen, read the manufacturer's fitting instructions very carefully and check that you have all the necessary brackets and fittings. Ensure that the mounting brackets are located correctly and bolted down securely. Note that some manufacturers use hose clamps to retain the mounting brackets; these should be discarded as they are convenient to use but not strong enough for the task. Stronger clamps should be substituted; car exhaust pipe clamps of suitable size would be a good alternative. Ensure that the front forks can turn through the full steering lock available without fouling the fairing. With many types of frame-mounted fairing the handlebars will have to be altered or a different type fitted and the steering lock will be restricted by stops provided with the fittings. Also check that the fairing does not foul the front wheel or mudguard, in any steering position, under full fork compression. Re-route any cables, brake pipes or electrical wiring which may snag on the fairing and take great care to protect all electrical connections, using insulating tape. If the manufacturer's instructions are followed carefully at every stage no serious problems should be encountered. Remember that hydraulic pipes that have been disconnected must be carefully re-tightened and the hydraulic system purged of air bubbles by bleeding.

Two things will become immediately apparent when taking a motorcycle on the road for the first time with a fairing – the first is the tendency to underestimate the road speed because of the lack of wind pressure on the body. This must be very carefully watched until one has grown accustomed to riding behind the fairing. The second thing is the alarming increase in engine noise which is an unfortunate but inevitable by-product of fitting any type of fairing or windscreen, and is caused by normal engine noise being reflected, and in some cases amplified, by the flat surface of the fairing.

## Luggage racks or carriers

Carriers are possibly the commonest item to be fitted to modern motorcycles. They vary enormously in size, carrying capacity, and durability. When selecting a carrier, always look for one which is made specifically for your machine and which is bolted on with as few separate brackets as possible. The universal-type carrier, with its mass of brackets and adaptor pieces, will generally prove too weak to be of any real use. A good carrier should bolt to the main frame, generally using the two suspension unit top mountings and a mudguard mounting bolt as attachment points, and have its luggage platform as low and as far forward as possible to minimise the effect of any load on the machine's stability. Look for good quality, heavy gauge tubing, good welding and good finish. Also ensure that the carrier does not prevent opening of the seat, sidepanels or tail compartment, as appropriate. When using a carrier, be very careful not to overload it. Excessive weight placed so high and so far to the rear of any motorcycle will have an adverse effect on the machine's steering and stability.

## Luggage

Motorcycle luggage can be grouped under two headings: soft and hard. Both types are available in many sizes and styles and have advantages and disadvantages in use.

Soft luggage is now becoming very popular because of its lower cost and its versatility. Whether in the form of tankbags, panniers, or strap-on bags, soft luggage requires in general no brackets and no modification to the motorcycle. Equipment can be swapped easily from one motorcycle to another and can be fitted and removed in seconds. Awkwardly shaped loads can easily be carried. The disadvantages of soft luggage are that the contents cannot be secure against the casual thief, very little protection is afforded in the event of a crash, and waterproofing is generally poor. Also, in the case of panniers, carrying capacity is restricted to approximately 10 lb, although this amount will vary considerably depending on the manufacturer's recommendation. When purchasing soft luggage, look for good quality material, generally vinyl or nylon, with strong, well-stitched attachment points. It is always useful to have separate pockets, especially on tank bags, for items which will be needed on the journey. When purchasing a tank bag, look for one which has a separate, well-padded, base. This will protect the tank's paintwork and permit easy access to the filler cap at petrol stations.

Hard luggage is confined to two types: panniers, and top boxes or tail trunks. Most hard luggage manufacturers produce matching sets of these items, the basis of which is generally that manufacturer's own heavy-duty luggage rack. Variations on this theme occur in the form of separate frames for the better quality panniers, fixed or quickly-detachable luggage, and in size and carrying capacity. Hard luggage offers a reasonable degree of security against theft and good protection against weather and accident damage. Carrying capacity is greater than that of soft luggage, around 15 – 20 lb in the case of panniers, although top boxes should never be loaded as much as their apparent capacity might imply. A top box should only be used for lightweight items, because one that is heavily laden can have a serious effect on the stability of the machine. When purchasing hard luggage look for the same good points as mentioned under fairings and windscreens, ie good quality mounting brackets and fittings, and well-finished fibreglass or ABS plastic cases. Again as with fairings, always purchase luggage made specifically for your motorcycle, using as few separate brackets as possible, to ensure that everything remains securely bolted in place. When fitting hard luggage, be careful to check that the rear suspension and brake operation will not be impaired in any way and remember that many pannier kits require re-siting of the indicators. Remember also that a non-standard exhaust system may make fitting extremely difficult.

## Handlebars

The occupation of fitting alternative types of handlebar is extremely popular with modern motorcyclists, whose motives may vary from the purely practical, wishing to improve the comfort of their machines, to the purely aesthetic, where form is more important than function. Whatever the reason, there are several considerations to be borne in mind when changing the handlebars of your machine. If fitting lower bars, check carefully that the switches and cables do not foul the petrol tank on full lock and that the surplus length of cable, brake pipe,

and electrical wiring are smoothly and tidily disposed of. Avoid tight kinks in cable or brake pipes which will produce stiff controls or the premature and disastrous failure of an overstressed component. If necessary, remove the petrol tank and re-route the cable from the engine/gearbox unit upwards, ensuring smooth gentle curves are produced. In extreme cases, it will be necessary to purchase a shorter brake pipe to overcome this problem. In the case of higher handlebars than standard it will almost certainly be necessary to purchase extended cables and brake pipes. Fortunately, many standard motorcycles have a custom version which will be equipped with higher handlebars and, therefore, factory-built extended components will be available from your local dealer. It is not usually necessary to extend electrical wiring, as switch clusters may be used on several different motorcycles, some being custom versions. This point should be borne in mind however when fitting extremely high or wide handlebars.

When fitting different types of handlebar, ensure that the mounting clamps are correctly tightened to the manufacturer's specifications and that cables and wiring, as previously mentioned, have smooth easy runs and do not snag on any part of the motorcycle throughout the full steering lock. Ensure that the fluid level in the front brake master cylinder remains level to avoid any chance of air entering the hydraulic system. Also check that the cables are adjusted correctly and that all handlebar controls operate correctly and can be easily reached when riding.

## Crashbars

Crashbars, also known as engine protector bars, engine guards, or case savers, are extremely useful items of equipment which can contribute protection to the machine's structure if a crash occurs. They do not, as has been inferred in the US, prevent the rider from crashing, or necessarily prevent rider injury should a crash occur.

It is recommended that only the smaller, neater, engine protector type of crashbar is considered. This type will offer protection while restricting, as little as is possible, access to the engine and the machine's ground clearance. The crashbars should be designed for use specifically on your machine, and should be constructed of heavy-gauge tubing with strong, integral mounting brackets. Where possible, they should bolt to a strong lug on the frame, usually at the engine mounting bolts.

The alternative type of crashbar is the larger cage type. This type is not recommended in spite of their appearance which promises some protection to the rider as well as to the machine. The larger amount of leverage imposed by the size of this type of crashbar increases the risk of severe frame damage in the event of an accident. This type also decreases the machine's ground clearance and restricts access to the engine. The amount of protection afforded the rider is open to some doubt as the design is based on the premise that the rider will stay in the normally seated position during an accident, and the crash bar structure will not itself fail. Neither result can in any way be guaranteed.

As a general rule, always purchase the best, ie usually the most expensive, set of crashbars you an afford. The investment will be repaid by minimising the amount of damage incurred, should the machine be involved in an accident. Finally, avoid the universal type of crashbar. This should be regarded only as a last resort to be used if no alternative exists. With its usual multitude of separate brackets and spacers, the universal crashbar is far too weak in design and construction to be of any practical value.

## Exhaust systems

The fitting of aftermarket exhaust systems is another extremely popular pastime amongst motorcyclists. The usual motive is to gain more performance from the engine but other considerations are to gain more ground clearance, to lose weight from the motorcycle, to obtain a more distinctive exhaust note or to find a cheaper alternative to the manufacturer's original equipment exhaust system. Original equipment exhaust systems often cost more and may well have a relatively short life. It should be noted that it is rare for an aftermarket exhaust system alone to give a noticeable increase in the engine's power output. Modern motorcycles are designed to give the highest power output possible allowing for factors such as quietness, fuel economy, spread of power, and long-term reliability. If there were a magic formula which allowed the exhaust system to produce more power without affecting these other considerations you can be sure that the manufacturers, with their large research and development facilities, would have found it and made use of it. Performance increases of a worthwhile and noticeable nature only come from well-tried and properly matched modifications to the entire engine, from the air filter, through the carburettors, port timing or camshaft and valve design, combustion chamber shape, compression ratio, and the exhaust system. Such modifications are well outside the scope of this manual but interested owners might refer to specialist books produced by the publisher of this manual which go into the whole subject in great detail.

Whatever your motive for wishing to fit an alternative exhaust system, be sure to seek expert advice before doing so. Changes to the carburettor jetting will almost certainly be required for which you must consult the exhaust system manufacturer. If he cannot supply adequately specific information it is reasonable to assume that insufficient development work has been carried out, and that particular make should be avoided. Other factors to be borne in mind are whether the exhaust system allows the use of both centre and side stands, whether it allows sufficient access to permit oil and filter changing and whether modifications are necessary to the standard exhaust system. Many two-stroke expansion chamber systems require the use of the standard exhaust pipe; this is all very well if the standard exhaust pipe and silencer are separate units but can cause problems if the two, as with so many modern two-strokes, are a one-piece unit. While the exhaust pipe can be removed easily by means of a hacksaw it is not so easy to refit the original silencer should you at any time wish to return the machine to standard trim. The same applies to several four-stroke systems.

On the subject of the finish of aftermarket exhausts, avoid black-painted systems unless you enjoy painting. As any trail-bike owner will tell you, rust has a great affinity for black exhausts and re-painting or rust removal becomes a task which must be carried out with monotonous regularity. A bright chrome finish is, as a general rule, a far better proposition as it is much easier to keep clean and to prevent rusting. Although the general finish of aftermarket exhaust systems is not always up to the standard of the original equipment the lower cost of such systems does at least reflect this fact.

When fitting an alternative system always purchase a full set of new exhaust gaskets, to prevent leaks. Fit the exhaust first to the cylinder head or barrel, as appropriate, tightening the retaining nuts or bolts by hand only and then line up the exhaust rear mountings. If the new system is a one-piece unit and the rear mountings do not line up exactly, spacers must be fabricated to take up the difference. Do not force the system into place as the stress thus imposed will rapidly cause cracks and splits to appear. Once all the mountings are loosely fixed, tighten the retaining nuts or bolts securely, being careful not to overtighten them. Where the motorcycle manufacturer's torque settings are available, these should be used. Do not forget to carry out any carburation changes recommended by the exhaust system's manufacturer.

## Electrical equipment

The vast range of electrical equipment available to motorcyclists is so large and so diverse that only the most general outline can be given here. Electrical accessories vary from electric ignition kits fitted to replace contact breaker points, to additional lighting at the front and rear, more powerful horns, various instruments and gauges, clocks, anti-theft systems, heated clothing, CB radios, radio-cassette players, and intercom systems, to name but a few of the more popular items of equipment.

As will be evident, it would require a separate manual to cover this subject alone and this section is therefore restricted to outlining a few basic rules which must be borne in mind when fitting electrical equipment. The first consideration is whether your machine's electrical system has enough reserve capacity to cope with the added demand of the accessories you wish to fit. The motorcycle's manufacturer or importer should be able to furnish this sort of information and may also be able to offer advice on uprating the electrical system. Failing this, a good dealer or the accessory manufacturer may be able to help. In some cases, more powerful generator components may be available, perhaps from another motorcycle in the manufacturer's range. The second consideration is the legal requirements in force in your area. The local police may be prepared to help with this point. In the UK for example, there are strict regulations governing the position and use of auxiliary riding lamps and fog lamps.

When fitting electrical equipment always disconnect the battery first to prevent the risk of a short-circuit, and be careful to ensure that all connections are properly made and that they are waterproof. Remember that many electrical accessories are designed primarily for use in cars and that they cannot easily withstand the exposure to vibration and to the weather. Delicate components must be rubber-mounted to insulate them from vibration, and sealed carefully to prevent the entry of rainwater and dirt. Be careful to follow exactly the accessory manufacturer's instructions in conjunction with the wiring diagram at the back of this manual.

## Accessories – general

Accessories fitted to your motorcycle will rapidly deteriorate if not cared for. Regular washing and polishing will maintain the finish and will provide an opportunity to check that all mounting bolts and nuts are securely fastened. Any signs of chafing or wear should be watched for, and the cause cured as soon as possible before serious damage occurs.

As a general rule, do not expect the re-sale value of your motorcycle to increase by an amount proportional to the amount of money and effort put into fitting accessories. It is usually the case that an absolutely standard motorcycle will sell more easily at a better price than one that has been modified. If you are in the habit of exchanging your machine for another at frequent intervals, this factor should be borne in mind to avoid loss of money.

# Fault diagnosis

**Contents**

## 1  Introduction

This Section provides an easy reference-guide to the more common faults that are likely to afflict your machine. Obviously, the opportunities are almost limitless for faults to develop as a result of obscure failures, and to try and cover all eventualities would require a book. Indeed, a number have been written on the subject.

Successful fault diagnosis is not a mysterious 'black art' but the application of a little knowledge combined with a systematic and logical approach to the problem. Approach any fault diagnosis by first accurately identifying the symptom and then checking through the list of possible causes, starting with the simplest or most obvious and progressing in stages to the most complex. Take nothing for granted, but above all apply liberal quantities of common sense.

The main symptom of a fault is given in the text as a major heading below which are listed, as Sections headings, the various systems or areas which may contain the fault. Details of each possible cause for a fault and the remedial action to be taken are given, in brief, in the paragraphs below each Section heading. Further information should be sought in the relevant Chapter.

### *Engine does not start when turned over*

## 2  No fuel flow to carburettor

● Fuel tank empty or level too low. Check that the tap is turned to 'On' or 'Reserve' position as required. If in doubt, prise off the fuel feed pipe at the carburettor end and check that fuel runs from pipe when the tap is turned on.

● Tank filler cap vent obstructed. This can prevent fuel from flowing into the carburettor float bowl because air cannot enter the fuel tank to replace it. The problem is more likely to appear when the machine is being ridden. Check by listening close to the filler cap and releasing it. A hissing noise indicates that a blockage is present. Remove the cap and clear the vent hole with wire or by using an air line from the inside of the cap.

● Fuel tap or filter blocked. Blockage may be due to accumulation of rust or paint flakes from the tank's inner surface or of foreign matter from contaminated fuel. Remove the tap and clean it and the filter. Look also for water droplets in the fuel.

● Fuel line blocked. Blockage of the fuel line is more likely to result from a kink in the line rather than the accumulation of debris.

## 3  Fuel not reaching cylinder

● Float chamber not filling. Caused by float needle or floats sticking in up position. This may occur after the machine has been left standing for an extended length of time allowing the fuel to evaporate. When this occurs a gummy residue is often left which hardens to a varnish-like substance. This condition may be worsened by corrosion and crystalline deposits produced prior to the total evaporation of contaminated fuel. Sticking of the float needle may also be caused by wear. In any case removal of the float chamber will be necessary for inspection and cleaning.

● Blockage in starting circuit, slow running circuit or jets. Blockage of these items may be attributable to debris from the fuel tank by-passing the filter system or to gumming up as described in paragraph 1. Water droplets in the fuel will also block jets and passages. The carburettor should be dismantled for cleaning.

● Fuel level too low. The fuel level in the float chamber is controlled by float height. The fuel level may increase with wear or damage but will never reduce, thus a low fuel level is an inherent rather than a developing condition. Check the float height, renewing the float or needle if required.

## 4  Engine flooding

● Float valve needle worn or stuck open. A piece of rust or other debris can prevent correct seating of the needle against the valve seat thereby permitting an uncontrolled flow of fuel. Similarly, a worn needle or needle seat will prevent valve closure. Dismantle the carburettor float bowl for cleaning and, if necessary, renewal of the worn components.

● Fuel level too high. The fuel level is controlled by the float height which may increase due to wear of the float needle, pivot pin or operating tang. Check the float height, and make any necessary adjustments. A leaking float will cause an increase in fuel level, and thus should be renewed.

● Cold starting mechanism. Check the choke (starter mechanism) for correct operation. If the mechanism jams in the 'On' position subsequent starting of a hot engine will be difficult.

● Blocked air filter. A badly restricted air filter will cause flooding. Check the filter and clean or renew as required. A collapsed inlet hose will have a similar effect. Check that the air filter inlet has not become blocked by a rag or similar item.

## 5  No spark at plug

● Ignition switch not on.
● Engine stop switch off.
● Spark plug dirty or fouled. Because the induction mixture of a two-stroke engine is inclined to be of a rather oily nature it is comparatively easy to foul the plug electrodes, especially where there have been repeated attempts to start the engine. A machine used for short journeys will be more prone to fouling because the engine may never reach full operating temperature, and the deposits will not burn off. On rare occasions a change of plug grade may be required but the advice of a dealer should be sought before making such a change. On all two-stroke machines it is a sound precaution to carry a new spare spark plug for substitution in the event of fouling problems.

● Spark plug failure. Clean the spark plug thoroughly and reset the electrode gap. Refer to the spark plug section and the colour condition guide in Routine maintenance. If the spark plug shorts internally or has sustained visible damage to the electrodes, core or ceramic insulator it should be renewed. On rare occasions a plug that appears to spark vigorously will fail to do so when refitted to the engine and subjected to the compression pressure in the cylinder.

● Spark plug cap or high tension (HT) lead faulty. Check condition and security. Replace if deterioration is evident. Spark plug caps have an internal resistor designed to inhibit electrical interference with radio and television sets. On rare occasions the resistor may break down, thus preventing sparking. If this is suspected, fit a new cap as a precaution.

● Spark plug cap loose. Check that the spark plug cap fits securely over the plug and, where fitted, the screwed terminal on the plug end is secure.

● Shorting due to moisture. Certain parts of the ignition system are susceptible to shorting when the machine is ridden or parked in wet weather. Check particularly the area from the spark plug cap back to the ignition coil. A water dispersant spray may be used to dry out waterlogged components. Recurrence of the problem can be prevented by using an ignition sealant spray after drying out and cleaning.

● Ignition or stop switch shorted. May be caused by water corrosion or wear. Water dispersant and contact cleaning sprays may be used. If this fails to overcome the problem dismantling and visual inspection of the switches will be required.

● Shorting or open circuit in wiring. Failure in any wire connecting any of the ignition components will cause ignition malfunction. Check also that all connections are clean, dry and tight.

● Ignition coil failure. Check the coil, referring to Chapter 3.

● Electronic ignition component failure, see Chapter 3.

## 6  Weak spark at plug

● Feeble sparking at the plug may be caused by any of the faults mentioned in the preceding Section other than those items in the first two paragraphs. Check first the spark plug, this being the most likely culprit.

## 7 Compression low

● Spark plug loose. This will be self-evident on inspection, and may be accompanied by a hissing noise when the engine is turned over. Remove the plug and check that the threads in the cylinder head are not damaged. Check also that the plug sealing washer is in good condition.

● Cylinder head gasket leaking. This condition is often accompanied by a high pitched squeak from around the cylinder head and oil loss, and may be caused by insufficiently tightened cylinder head fasteners, a warped cylinder head or mechanical failure of the gasket material. Re-torqueing the fasteners to the correct specification may seal the leak in some instances but if damage has occurred this course of action will provide, at best, only a temporary cure.

● Low crankcase compression. This can be caused by worn main bearings and seals and will upset the incoming fuel/air mixture. A good seal in these areas is essential on any two-stroke engine.

● Piston rings sticking or broken. Sticking of the piston rings may be caused by seizure due to lack of lubrication or overheating as a result of poor carburation or incorrect fuel type. Gumming of the rings may result from lack of use, or carbon deposits in the ring grooves. Broken rings result from over-revving, over-heating or general wear. In either case a top-end overhaul will be required.

## Engine stalls after starting

### 8 General causes

● Improper cold start mechanism operation. Check that the operating controls function smoothly and, where applicable, are correctly adjusted. A cold engine may not require application of an enriched mixture to start initially but may baulk without choke once firing. Likewise a hot engine may start with an enriched mixture but will stop almost immediately if the choke is inadvertently in operation.

● Ignition malfunction. See Section 9. Weak spark at plug.

● Carburettor incorrectly adjusted. Maladjustment of the mixture strength or idle speed may cause the engine to stop immediately after starting. See Routine maintenance.

● Fuel contamination. Check for filter blockage by debris or water which reduces, but does not completely stop, fuel flow, or blockage of the slow speed circuit in the carburettor by the same agents. If water is present it can often be seen as droplets in the bottom of the float bowl. Clean the filter and, where water is in evidence, drain and flush the fuel tank and float bowl.

● Intake air leak. Check for security of the carburettor mounting and hose connections, and for cracks or splits in the hoses. Check also that the carburettor top is secure.

● Air filter blocked or omitted. A blocked filter will cause an over-rich mixture; the omission of a filter will cause an excessively weak mixture. Both conditions will have a detrimental effect on carburation. Clean or renew the filter as necessary.

● Fuel filler cap air vent blocked. Usually caused by dirt or water. Clean the vent orifice.

● Choked exhaust system. Caused by excessive carbon build-up in the system, particularly around the silencer baffles. Refer to Routine maintenance for further information.

● Excessive carbon build-up in the engine. This can result from failure to decarbonise the engine at the specified interval or through excessive oil consumption. Check pump adjustment.

## Poor running at idle and low speed

### 9 Weak spark at plug or erratic firing

● Spark plug fouled, faulty or incorrectly adjusted. See Section 5 or refer to Routine maintenance.

● Spark plug cap or high tension lead shorting. Check the condition of both these items ensuring that they are in good condition and dry and that the cap is fitted correctly.

● Spark plug type incorrect. Fit plug of correct type and heat range as given in Specifications. In certain conditions a plug of hotter or colder type may be required for normal running.

● Ignition timing incorrect. Check the ignition timing and ensure that the advance is functioning correctly.

● Faulty ignition coil. Partial failure of the coil internal insulation will diminish the performance of the coil. No repair is possible, a new component must be fitted.

● Defective flywheel generator ignition source coil. Refer to Chapter 3 for further details on test procedures.

## 10 Fuel/air mixture incorrect

● Intake air leak. Check carburettor mountings and air cleaner hoses for security and signs of splitting. Ensure that carburettor top is tight.

● Mixture strength incorrect. Adjust slow running mixture strength using pilot adjustment screw (where fitted).

● Pilot jet or slow running circuit blocked. The carburettor should be removed and dismantled for thorough cleaning. Blow through all jets and air passages with compressed air to clear obstructions.

● Air cleaner clogged or omitted. Clean or fit air cleaner element as necessary. Check also that the element and air filter cover are correctly seated.

● Cold start mechanism in operation. Check that the choke has not been left on inadvertently and the operation is correct.

● Fuel level too high or too low. Check the float height, renewing float or needle if required. See Section 3 or 4.

● Fuel tank air vent obstructed. Obstructions usually caused by dirt or water. Clean vent orifice.

## 11 Compression low

● See Section 7.

## Acceleration poor

### 12 General causes

● All items as for previous Section.

● Choked air filter. Failure to keep the air filter element clean will allow the build-up of dirt with proportional loss of performance. In extreme cases of neglect acceleration will suffer.

● Choked exhaust system. This can result from failure to remove accumulations of carbon from the silencer baffles at the prescribed intervals. The increased back pressure will make the machine noticeably sluggish. Refer to Routine maintenance for further information on decarbonisation.

● Excessive carbon build-up in the engine. This can result from failure to decarbonise the engine at the specified interval or through excessive oil consumption. Check pump adjustment.

● Ignition timing incorrect. Check the ignition timing as described in Routine maintenance.

● Carburation fault. See Section 10.

● Mechanical resistance. Check that the brakes are not binding. On small machines in particular note that the increased rolling resistance caused by under-inflated tyres may impede acceleration.

## Poor running or lack of power at high speeds

### 13 Weak spark at plug or erratic firing

● All items as for Section 9.
● HT lead insulation failure. Insulation failure of the HT lead and spark plug cap due to old age or damage can cause shorting when the engine is driven hard. This condition may be less noticeable, or not noticeable at all at lower engine speeds.

### 14 Fuel/air mixture incorrect

● All items as for Section 10, with the exception of items relative exclusively to low speed running.
● Main jet blocked. Debris from contaminated fuel, or from the fuel tank, and water in the fuel can block the main jet. Clean the fuel filter, the float bowl area, and if water is present, flush and refill the fuel tank.
● Main jet is the wrong size. The standard carburettor jetting is for sea level atmospheric pressure. For high altitudes, usually above 5000 ft, a smaller main jet will be required.
● Jet needle and needle jet worn. These can be renewed individually but should be renewed as a pair. Renewal of both items requires complete dismantling of the carburettor.
● Air bleed holes blocked. Dismantle carburettor and use compressed air to blow out all air passages.
● Reduced fuel flow. A reduction in the maximum fuel flow from the fuel tank to the carburettor will cause fuel starvation, proportionate to the engine speed. Check for blockages through debris or a kinked fuel line.

### 15 Compression low

● See Section 7.

## Knocking or pinking

### 16 General causes

● Carbon build-up in combustion chamber. After high mileages have been covered large accumulations of carbon may occur. This may glow red hot and cause premature ignition of the fuel/air mixture, in advance of normal firing by the spark plug. Cylinder head removal will be required to allow inspection and cleaning.
● Fuel incorrect. A low grade fuel, or one of poor quality may result in compression induced detonation of the fuel resulting in knocking and pinking noises. Old fuel can cause similar problems. A too highly leaded fuel will reduce detonation but will accelerate deposit formation in the combustion chamber and may lead to early pre-ignition as described in item 1.
● Spark plug heat range incorrect. Uncontrolled pre-ignition can result from the use of a spark plug the heat range of which is too hot.
● Weak mixture. Overheating of the engine due to a weak mixture can result in pre-ignition occurring where it would not occur when engine temperature was within normal limits. Maladjustment, blocked jets or passages and air leaks can cause this condition.

## Overheating

### 17 Firing incorrect

● Spark plug fouled, defective or maladjusted. See Section 5.
● Spark plug type incorrect. Refer to the Specifications and ensure that the correct plug type is fitted.
● Incorrect ignition timing. Timing that is far too much advanced or far too much retarded will cause overheating. Check the ignition timing is correct.

### 18 Fuel/air mixture incorrect

● Slow speed mixture strength incorrect. Adjust pilot screw (where fitted).
● Main jet wrong size. The carburettor is jetted for sea level atmospheric conditions. For high altitudes, usually above 5000 ft, a smaller main jet will be required.
● Air filter badly fitted or omitted. Check that the filter element is in place and that it and the air filter box cover are sealing correctly. Any leaks will cause a weak mixture.
● Induction air leaks. Check the security of the carburettor mountings and hose connections, and for cracks and splits in the hoses. Check also that the carburettor top is secure.
● Fuel level too low. See Section 3.
● Fuel tank filler cap air vent obstructed. Clear blockage.

### 19 Lubrication inadequate

● Oil pump settings incorrect. The oil pump settings are of great importance since the quantities of oil being injected are very small. Any variation in oil delivery will have a significant effect on the engine. Refer to Routine maintenance for further information.
● Oil tank empty or low. This will have disastrous consequences if left unnoticed. Check and replenish tank regularly.
● Transmission oil low or worn out. Check the level regularly and investigate any loss of oil. If the oil level drops with no sign of external leakage it is likely that the crankshaft main bearing oil seals are worn, allowing transmission oil to be drawn into the crankcase during induction.

### 20 Miscellaneous causes

● Engine fins clogged. A build-up of mud in the cylinder head and cylinder barrel cooling fins will decrease their cooling capabilities. Clean the fins as required.

## Clutch operating problems

### 21 Clutch slip

● No clutch lever play. Adjust clutch lever end play according to the procedure in Routine maintenance.
● Friction plates worn or warped. Overhaul clutch assembly, replacing plates out of specification.
● Steel plates worn or warped. Overhaul clutch assembly, replacing plates out of specification.
● Clutch springs broken or worn. Old or heat-damaged (from slipping clutch) springs should be replaced with new ones.
● Clutch inner cable snagging. Caused by a frayed cable or kinked outer cable. Replace the cable with a new one. Repair of a frayed cable is not advised.
● Clutch release mechanism defective. Worn or damaged parts in the clutch release mechanism could include the release lever or centre piece. Replace parts as necessary.
● Clutch hub and outer drum worn. Severe indentation by the clutch plate tangs of the channels in the hub and drum will cause snagging of the plates preventing correct engagement. If this damage occurs, renewal of the worn components is required.
● Lubricant incorrect. Use of a transmission lubricant other than that specified may allow the plates to slip.

### 22 Clutch drag

● Clutch lever play excessive. Adjust lever at bars or at cable end if necessary.
● Clutch plates warped or damaged. This will cause a drag on the clutch, causing the machine to creep. Overhaul clutch assembly.

● Clutch spring tension uneven. Usually caused by a sagged or broken spring. Check and replace springs.
● Transmission oil deteriorated. Badly contaminated transmission oil and a heavy deposit of oil sludge on the plates will cause plate sticking. The oil recommended for this machine is of the detergent type, therefore it is unlikely that this problem will arise unless regular oil changes are neglected.
● Transmission oil viscosity too high. Drag in the plates will result from the use of an oil with too high a viscosity. In very cold weather clutch drag may occur until the engine has reached operating temperature.
● Clutch hub and outer drum worn. Indentation by the clutch plate tangs of the channels in the hub and drum will prevent easy plate disengagement. If the damage is light the affected areas may be dressed with a fine file. More pronounced damage will necessitate renewal of the components.
● Clutch housing seized to shaft. Lack of lubrication, severe wear or damage can cause the housing to seize to the shaft. Overhaul of the clutch, and perhaps the transmission, may be necessary to repair damage.
● Clutch release mechanism defective. Worn or damaged release mechanism parts can stick and fail to provide leverage. Overhaul clutch cover components.
● Loose clutch hub bolt. Causes drum and hub misalignment, putting a drag on the engine. Engagement adjustment continually varies. Overhaul clutch assembly.

## Gear selection problems

### 23 Gear lever does not return

● Weak or broken return spring. Renew the spring.
● Gearchange shaft bent or seized. Distortion of the gearchange shaft often occurs if the machine is dropped heavily on the gear lever. Provided that damage is not severe straightening of the shaft is permissible.

### 24 Gear selection difficult or impossible

● Clutch not disengaging fully. See Section 22.
● Gearchange shaft bent. This often occurs if the machine is dropped heavily on the gear lever. Straightening of the shaft is permissible if the damage is not too great.
● Gearchange arms, pawls or pins worn or damaged. Wear or breakage of any of these items may cause difficulty in selecting one or more gears. Overhaul the selector mechanism.
● Gearchange drum set levers damaged. Failure, rather than wear of these items may jam the drum thereby preventing gearchanging or causing false selection at high speed.
● Selector forks bent or seized. This can be caused by dropping the machine heavily on the gearchange lever or as a result of lack of lubrication. Though rare, bending of a shaft can result from a missed gearchange or false selection at high speed.
● Selector fork end and pin wear. Pronounced wear of these items and the grooves in the gearchange drum can lead to imprecise selection and, eventually, no selection. Renewal of the worn components will be required.
● Structural failure. Failure of any one component of the selector rod and change mechanism will result in improper or fouled gear selection.

### 25 Jumping out of gear

● Set lever assembly worn or damaged. Wear of the lever and the notch with which it locates and breakage of the detent spring can cause imprecise gear selection resulting in jumping out of gear. Renew the damaged components.
● Gear pinion dogs worn or damaged. Rounding off of the dog edges and the mating recesses in adjacent pinion can lead to jumping out of gear when under load. The gears should be inspected and renewed. Attempting to reprofile the dogs is not recommended.
● Selector forks, gearchange drum and pinion grooves worn. Extreme wear of these interconnected items can occur after high mileages especially when lubrication has been neglected. The worn components must be renewed.
● Gear pinions, bushes and shafts worn. Renew the worn components.
● Bent gearchange shaft. Often caused by dropping the machine on the gear lever.
● Gear pinion tooth broken. Chipped teeth are unlikely to cause jumping out of gear once the gear has been selected fully; a tooth which is completely broken off, however, may cause problems in this respect and in any event will cause transmission noise.

### 26 Overselection

● Pawl spring weak or broken. Renew the spring.
● Set lever worn or broken. Renew the damaged items.
● Set lever spring worn or broken. Renew the spring.
● Gearchange arm stop pads worn. Repairs can be made by welding and reprofiling with a file.

## Abnormal engine noise

### 27 Knocking or pinking

● See Section 16.

### 28 Piston slap or rattling from cylinder

● Cylinder bore/piston clearance excessive. Resulting from wear, or partial seizure. This condition can often be heard as a high, rapid tapping noise when the engine is under little or no load, particularly when power is just beginning to be applied. Reboring to the next correct oversize should be carried out (where possible) and a new oversize piston fitted.
● Connecting rod bent. This can be caused by over-revving, trying to start a very badly flooded engine (resulting in a hydraulic lock in the cylinder) or by earlier mechanical failure. Attempts at straightening a bent connecting rod are not recommended. Careful inspection of the crankshaft should be made before renewing the damaged connecting rod.
● Gudgeon pin, piston boss bore or small-end bearing wear or seizure. Excess clearance or partial seizure between normal moving parts of these items can cause continuous or intermittent tapping noises. Rapid wear or seizure is caused by lubrication starvation.
● Piston rings worn, broken or sticking. Renew the rings after careful inspection of the piston and bore.

### 29 Other noises

● Big-end bearing wear. A pronounced knock from within the crankcase which worsens rapidly is indicative of big-end bearing failure as a result of extreme normal wear or lubrication failure. Remedial action in the form of a bottom end overhaul should be taken; continuing to run the engine will lead to further damage including the possibility of connecting rod breakage.
● Main bearing failure. Extreme normal wear or failure of the main bearings is characteristically accompanied by a rumble from the crankcase and vibration felt through the frame and footrests. Renew the worn bearings and carry out a very careful examination of the crankshaft.
● Crankshaft excessively out of true. A bent crank may result from over-revving or damage from an upper cylinder component or gearbox failure. Damage can also result from dropping the machine on either crankshaft end. Straightening of the crankshaft is not possible in normal circumstances; a replacement item should be fitted.
● Engine mounting loose. Tighten all the engine mounting nuts and bolts.
● Cylinder head gasket leaking. The noise most often associated with a leaking head gasket is a high pitched squeaking, although any

other noise consistent with gas being forced out under pressure from a small orifice can also be emitted. Gasket leakage is often accompanied by oil seepage from around the mating joint or from the cylinder head holding down bolts and nuts. Leakage results from insufficient or uneven tightening of the cylinder head fasteners, or from random mechanical failure. Retightening to the correct torque figure will, at best, only provide a temporary cure. The gasket should be renewed at the earliest opportunity.

● Exhaust system leakage. Popping or crackling in the exhaust system, particularly when it occurs with the engine on the overrun, indicates a poor joint either at the cylinder port or at the exhaust pipe/silencer connection. Failure of the gasket or looseness of the clamp should be looked for.

## Abnormal transmission noise

### 30 Clutch noise

● Clutch outer drum/friction plate tang clearance excessive.
● Clutch outer drum/spacer clearance excessive.
● Clutch outer drum/thrust washer clearance excessive.
● Primary drive gear teeth worn or damaged.
● Clutch shock absorber assembly worn or damaged.

### 31 Transmission noise

● Bearing or bushes worn or damaged. Renew the affected components.
● Gear pinions worn or chipped. Renew the gear pinions.
● Metal chips jammed in gear teeth. This can occur when pieces of metal from any failed component are picked up by a meshing pinion. The condition will lead to rapid bearing wear or early gear failure.
● Gearbox oil level too low. Top up immediately to prevent damage to gearbox.
● Gearchange mechanism worn or damaged. Wear or failure of certain items in the selection and change components can induce mis-selection of gears (see Section 24) where incipient engagement of more than one gear set is promoted. Remedial action, by the overhaul of the gearbox, should be taken without delay.
● Chain snagging on cases or cycle parts. A badly worn chain or one that is excessively loose may snag or smack against adjacent components.

## Exhaust smokes excessively

### 32 White/blue smoke (caused by oil burning)

● Oil pump settings incorrect. Check and reset the oil pump as described in Chapter 2.
● Crankshaft main bearing oil seals worn. Wear in the main bearing oil seals, often in conjunction with wear in the bearings themselves, can allow transmission oil to find its way into the crankcase and thence to the combustion chamber. This condition is often indicated by a mysterious drop in the transmission oil level with no sign of external leakage.
● Accumulated oil deposits in exhaust system. If the machine is used for short journeys only it is possible for the oil residue in the exhaust gases to condense in the relatively cool silencer. If the machine is then taken for a longer run in hot weather, the accumulated oil will burn off producing ominous smoke from the exhaust.

### 33 Black smoke (caused by over-rich mixture)

● Air filter element clogged. Clean or renew the element.
● Main jet loose or too large. Remove the float chamber to check for tightness of the jet. If the machine is used at high altitudes rejetting will be required to compensate for the lower atmospheric pressure.
● Cold start mechanism jammed on. Check that the mechanism works smoothly and correctly.
● Fuel level too high. The fuel level is controlled by the float height which can increase as a result of wear or damage. Remove the float bowl and check the float height. Check also that floats have not punctured; a punctured float will lose buoyancy and allow an increased fuel level.
● Float valve needle stuck open. Caused by dirt or a worn valve. Clean the float chamber or renew the needle and, if necessary, the valve seat.

## Poor handling or roadholding

### 34 Directional instability

● Steering head bearing adjustment too tight. This will cause rolling or weaving at low speeds. Re-adjust the bearings.
● Steering head bearing worn or damaged. Correct adjustment of the bearing will prove impossible to achieve if wear or damage has occurred. Inconsistent handling will occur including rolling or weaving at low speed and poor directional control at indeterminate higher speeds. The steering head bearing should be dismantled for inspection and renewed if required. Lubrication should also be carried out.
● Bearing races pitted or dented. Impact damage caused, perhaps, by an accident or riding over a pot-hole can cause indentation of the bearing, usually in one position. This should be noted as notchiness when the handlebars are turned. Renew and lubricate the bearings.
● Steering stem bent. This will occur only if the machine is subjected to a high impact such as hitting a curb or a pot-hole. The lower yoke/stem should be renewed; do not attempt to straighten the stem.
● Front or rear tyre pressures too low.
● Front or rear tyre worn. General instability, high speed wobbles and skipping over white lines indicates that tyre renewal may be required. Tyre induced problems, in some machine/tyre combinations, can occur even when the tyre in question is by no means fully worn.
● Swinging arm bearings worn. Difficulties in holding line, particularly when cornering or when changing power settings indicates wear in the swinging arm bearings. The swinging arm should be removed from the machine and the bearings renewed.
● Swinging arm flexing. The symptoms given in the preceding paragraph will also occur if the swinging arm fork flexes badly. This can be caused by structural weakness as a result of corrosion, fatigue or impact damage, or because the rear wheel spindle is slack.
● Wheel bearings worn. Renew the worn bearings.
● Loose wheel spokes. The spokes should be tightened evenly to maintain tension and trueness of the rim.
● Tyres unsuitable for machine. Not all available tyres will suit the characteristics of the frame and suspension, indeed, some tyres or tyre combinations may cause a transformation in the handling characteristics. If handling problems occur immediately after changing to a new tyre type or make, revert to the original tyres to see whether an improvement can be noted. In some instances a change to what are, in fact, suitable tyres may give rise to handling deficiences. In this case a thorough check should be made of all frame and suspension items which affect stability.

### 35 Steering bias to left or right

● Rear wheel out of alignment. Caused by uneven adjustment of chain tensioner adjusters allowing the wheel to be askew in the fork ends. A bent rear wheel spindle will also misalign the wheel in the swinging arm.

● Wheels out of alignment. This can be caused by impact damage to the frame, swinging arm, wheel spindles or front forks. Although occasionally a result of material failure or corrosion it is usually as a result of a crash.

● Front forks twisted in the steering yokes. A light impact, for instance with a pot-hole or low curb, can twist the fork logs in the steering yokes without causing structural damage to the fork legs or the yokes themselves. Re-alignment can be made by loosening the yoke pinch bolts, wheel spindle and mudguard bolts. Re-align the wheel with the handlebars and tighten the bolts working upwards from the wheel spindle. This action should be carried out only when there is no chance that structural damage has occurred.

### 36 Handlebar vibrates or oscillates

● Tyres worn or out of balance. Either condition, particularly in the front tyre, will promote shaking of the fork assembly and thus the handlebars. A sudden onset of shaking can result if a balance weight is displaced during use.

● Tyres badly positioned on the wheel rims. A moulded line on each wall of a tyre is provided to allow visual verification that the tyre is correctly positioned on the rim. A check can be made by rotating the tyre; any misalignment will be immediately obvious.

● Wheel rims warped or damaged. Inspect the wheels for runout as described in Chapter 5.

● Swinging arm bearings worn. Renew the bearings.

● Wheel bearings worn. Renew the bearings.

● Steering head bearings incorrectly adjusted. Vibration is more likely to result from bearings which are too loose rather than too tight. Re-adjust the bearings.

● Loose fork component fasteners. Loose nuts and bolts holding the fork legs, wheel spindle, mudguards or steering stem can promote shaking at the handlebars. Fasteners on running gear such as the forks and suspension should be check tightened occasionally to prevent dangerous looseness of components occurring.

● Engine mounting bolts loose. Tighten all fasteners.

### 37 Poor front fork performance

● Damping fluid level incorrect. If the fluid level is too low poor suspension control will occur resulting in a general impairment of roadholding and early loss of tyre adhesion when cornering and braking. Too much oil is unlikely to change the fork characteristics unless severe overfilling occurs when the fork action will become stiffer and oil seal failure may occur.

● Damping oil viscosity incorrect. The damping action of the fork is directly related to the viscosity of the damping oil. The lighter the oil used, the less will be the damping action imparted. For general use, use the recommended viscosity of oil, changing to a slightly higher or heavier oil only when a change in damping characteristic is required. Overworked oil, or oil contaminated with water which has found its way past the seals, should be renewed to restore the correct damping performance and to prevent bottoming of the forks.

● Damping components worn or corroded. Advanced normal wear of the fork internals is unlikely to occur until a very high mileage has been covered. Continual use of the machine with damaged oil seals which allows the ingress of water, or neglect, will lead to rapid corrosion and wear. Dismantle the forks for inspection and overhaul.

● Weak fork springs. Progressive fatigue of the fork springs, resulting in a reduced spring free length, will occur after extensive use. This condition will promote excessive fork dive under braking, and in its advanced form will reduce the at-rest extended length of the forks and thus the fork geometry. Renewal of the springs as a pair is the only satisfactory course of action.

● Bent stanchions or corroded stanchions. Both conditions will prevent correct telescoping of the fork legs, and in an advanced state can cause sticking of the fork in one position. In a mild form corrosion will cause stiction of the fork thereby increasing the time the suspension takes to react to an uneven road surface. Bent fork stanchions should be attended to immediately because they indicate that impact damage has occurred, and there is a danger that the forks will fail with disastrous consequences.

### 38 Front fork judder when braking (see also Section 50)

● Wear between the fork stanchions and the fork legs. Renewal of the affected components is required.

● Slack steering head bearings. Re-adjust the bearings.

● Warped brake disc or drum. If irregular braking action occurs fork judder can be induced in what are normally serviceable forks. Renew the damaged brake components.

### 39 Poor rear suspension performance

● Rear suspension unit damper worn out or leaking. The damping performance of most rear suspension units falls off with age. This is a gradual process, and thus may not be immediately obvious. Indications of poor damping include hopping of the rear end when cornering or braking, and a general loss of positive stability.

● Weak rear spring. If the suspension unit spring fatigues it will promote excessive pitching of the machine and reduce the ground clearance when cornering.

● Swinging arm flexing or bearings worn. See Sections 34 and 36.

● Bent suspension unit damper rod. This is likely to occur only if the machine is dropped or if seizure of the piston occurs.

## Abnormal frame and suspension noise

### 40 Front end noise

● Oil level low or too thin. This can cause a 'spurting' sound and is usually accompanied by irregular fork action.

● Spring weak or broken. Makes a clicking or scraping sound. Fork oil will have a lot of metal particles in it.

● Steering head bearings loose or damaged. Clicks when braking. Check, adjust or replace.

● Fork clamps loose. Make sure all fork clamp pinch bolts are tight.

● Fork stanchion bent. Good possibility if machine has been dropped. Repair or replace tube.

### 41 Rear suspension noise

● Fluid level too low. Leakage of suspension unit, usually evident by oil on the outer surfaces, can cause a spurting noise.

● Defective rear suspension unit with internal damage.

● Worn suspension linkage bearings. See Chapter 4.

## Brake problems

### 42 Brakes are spongy or ineffective – disc brakes

● Air in brake circuit. This is only likely to happen in service due to neglect in checking the fluid level or because a leak has developed. The problem should be identified and the brake system bled of air.

● Pad worn. Check the pad wear against the wear limits specified and renew the pads if necessary.

● Contaminated pads. Cleaning pads which have been contaminated with oil, grease or brake fluid is unlikely to prove successful; the pads should be renewed.

● Pads glazed. This is usually caused by overheating. The surface of the pads may be roughened using glass-paper or a fine file.

● Brake fluid deterioration. A brake which on initial operation is firm but rapidly becomes spongy in use may be failing due to water contamination of the fluid. The fluid should be drained and then the system refilled and bled.

● Master cylinder seal failure. Wear or damage of master cylinder internal parts will prevent pressurisation of the brake fluid. Overhaul the master cylinder unit.

● Caliper seal failure. This will almost certainly be obvious by loss of fluid, a lowering of fluid in the master cylinder reservoir and contamination of the brake pads and caliper. Overhaul the caliper assembly.

## 43 Brakes drag – disc brakes

● Disc warped. The disc must be renewed.
● Caliper piston, caliper or pads corroded. The brake caliper assembly is vulnerable to corrosion due to water and dirt, and unless cleaned at regular intervals and lubricated in the recommended manner, will become sticky in operation.
● Piston seal deteriorated. The seal is designed to return the piston in the caliper to the retracted position when the brake is released. Wear or old age can affect this function. The caliper should be overhauled if this occurs.
● Brake pad damaged. Pad material separating from the backing plate due to wear or faulty manufacture. Renew the pads. Faulty installation of a pad also will cause dragging.
● Wheel spindle bent. The spindle may be straightened if no structural damage has occurred.
● Brake lever or pedal not returning. Check that the lever or pedal works smoothly throughout its operating range and does not snag on any adjacent cycle parts. Lubricate the pivot if necessary.
● Twisted caliper support bracket. This is likely to occur only after impact in an accident. No attempt should be made to re-align the caliper; the bracket should be renewed.

## 44 Brake lever or pedal pulsates in operation – disc brakes

● Disc warped or irregularly worn. The disc must be renewed.
● Wheel spindle bent. The spindle may be straightened provided no structural damage has occurred.

## 45 Disc brake noise

● Brake squeal. This can be caused by dust on the pads, usually in combination with glazed pads, or other contamination from oil, grease, brake fluid or corrosion. Persistent squealing which cannot be traced to any of the normal causes can often be cured by applying a thin layer of high temperature silicone to the rear of the pads. Make absolutely certain that no grease is allowed to contaminate the braking surface of the pads.
● Glazed pads. This is usually caused by high temperatures or contamination. The pad surfaces may be roughened using glass-paper or a fine file. If this approach does not effect a cure the pads should be renewed.
● Disc warped. This can cause a chattering, clicking or intermittent squeal and is usually accompanied by a pulsating brake lever or pedal or uneven braking. The disc must be renewed.
● Brake pads fitted incorrectly or undersize. Longitudinal play in the pads due to omission of the locating springs (where fitted) or because pads of the wrong size have been fitted will cause a single tapping noise every time the brake is operated. Inspect the pads for correct installation and security.

## 46 Brakes are spongy or ineffective – drum brakes

● Brake cable deterioration. Damage to the outer cable by stretching or being trapped will give a spongy feel to the brake lever. The cable should be renewed. A cable which has become corroded due to old age or neglect of lubrication will partially seize making operation very heavy. Lubrication at this stage may overcome the problem but the fitting of a new cable is recommended.
● Worn brake linings. Determine lining wear using the external

brake wear indicator on the brake backplate, or by removing the wheel and withdrawing the brake backplate. Renew the shoe/lining units as a pair if the linings are worn below the recommended limit.
● Worn brake camshaft. Wear between the camshaft and the bearing surface will reduce brake feel and reduce operating efficiency. Renewal of one or both items will be required to rectify the fault.
● Worn brake cam and shoe ends. Renew the worn components.
● Linings contaminated with dust or grease. Any accumulations of dust should be cleaned from the brake assembly and drum using a petrol dampened cloth. Do not blow or brush off the dust because it is asbestos based and thus harmful if inhaled. Light contamination from grease can be removed from the surface of the brake linings using a solvent; attempts at removing heavier contamination are less likely to be successful because some of the lubricant will have been absorbed by the lining material which will severely reduce the braking performance.

## 47 Brake drag – drum brakes

● Incorrect adjustment. Re-adjust the brake operating mechanism.
● Drum warped or oval. This can result from overheating or impact or uneven tension of the wheel spokes. The condition is difficult to correct, although if slight ovality only occurs, skimming the surface of the brake drum can provide a cure. This is work for a specialist engineer. Renewal of the complete wheel hub is normally the only satisfactory solution.
● Weak brake shoe return springs. This will prevent the brake lining/shoe units from pulling away from the drum surface once the brake is released. The springs should be renewed.
● Brake camshaft, lever pivot or cable poorly lubricated. Failure to attend to regular lubrication of these areas will increase operating resistance which, when compounded, may cause tardy operation and poor release movement.

## 48 Brake lever or pedal pulsates in operation – drum brakes

● Drums warped or oval. This can result from overheating or impact or uneven spoke tension. This condition is difficult to correct, although if slight ovality only occurs skimming the surface of the drum can provide a cure. This is work for a specialist engineer. Renewal of the hub is normally the only satisfactory solution.

## 49 Drum brake noise

● Drum warped or oval. This can cause intermittent rubbing of the brake linings against the drum. See the preceding Section.
● Brake linings glazed. This condition, usually accompanied by heavy lining dust contamination, often induces brake squeal. The surface of the linings may be roughened using glass-paper or a fine file.

## 50 Brake induced fork judder

● Worn front fork stanchions and legs, or worn or badly adjusted steering head bearings. These conditions, combined with uneven or pulsating braking as described in Sections 44 and 48 will induce more or less judder when the brakes are applied, dependent on the degree of wear and poor brake operation. Attention should be given to both areas of malfunction. See the relevant Sections.

## *Electrical problems*

## 51 Battery dead or weak

● Battery faulty. Battery life should not be expected to exceed 3 to 4 years. Gradual sulphation of the plates and sediment deposits will reduce the battery performance. Plate and insulator damage can often occur as a result of vibration. Complete power failure, or intermittent failure, may be due to a broken battery terminal. Lack of electrolyte will prevent the battery maintaining charge.

● Battery leads making poor contact. Remove the battery leads and clean them and the terminals, removing all traces of corrosion and tarnish. Reconnect the leads and apply a coating of petroleum jelly to the terminals.

● Load excessive. If additional items such as spot lamps, are fitted, which increase the total electrical load above the maximum alternator output, the battery will fail to maintain full charge. Reduce the electrical load to suit the electrical capacity.

● Rectifier or ballast resistor failure.

● Alternator generating coils open-circuit or shorted.

● Charging circuit shorting or open circuit. This may be caused by frayed or broken wiring, dirty connectors or a faulty ignition switch. The system should be tested in a logical manner. See Section 54.

## 52 Battery overcharged

● Battery wrongly matched to the electrical circuit. Ensure that the specified battery is fitted to the machine.

## 53 Total electrical failure

● Fuse blown. Check the main fuse. If a fault has occurred, it must be rectified before a new fuse is fitted.

● Battery faulty. See Section 51.

● Earth failure. Check that the frame main earth strap from the battery is securely affixed to the frame and is making a good contact.

● Ignition switch or power circuit failure. Check for current flow through the battery positive lead (white) to the ignition switch. Check the ignition switch for continuity.

## 54 Circuit failure

● Cable failure. Refer to the machine's wiring diagram and check the circuit for continuity. Open circuits are a result of loose or corroded connections, either at terminals or in-line connectors, or because of broken wires. Occasionally, the core of a wire will break without there being any apparent damage to the outer plastic cover.

● Switch failure. All switches may be checked for continuity in each switch position, after referring to the switch position boxes incorporated in the wiring diagram for the machine. Switch failure may be a result of mechanical breakage, corrosion or water.

● Fuse blown. Refer to the wiring diagram to check whether or not a circuit fuse is fitted. Replace the fuse, if blown, only after the fault has been identified and rectified.

## 55 Bulbs blowing repeatedly

● Vibration failure. This is often an inherent fault related to the natural vibration characteristics of the engine and frame and is, thus, difficult to resolve. Modifications of the lamp mounting, to change the damping characteristics, may help.

● Intermittent earth. Repeated failure of one bulb, particularly where the bulb is fed directly from the generator, indicates that a poor earth exists somewhere in the circuit. Check that a good contact is available at each earthing point in the circuit.

# KAWASAKI AE/AR 50 & 80

## Check list

**Daily**

1. Top up the engine oil tank if the oil light stays on with the engine running and in gear
2. Check the disc brake hydraulic fluid level through the reservoir sight glass
3. Check the brake stop lamp is functioning properly

**Weekly, or every 125 miles (200 km)**

1. Check around the machine for loose fasteners, frayed cables and oil and petrol leaks
2. Ensure the lights and horn are functioning properly
3. Check the gearbox oil level
4. Lubricate exposed sections of control cables
5. Check the tyre pressures and check the tyres for wear and damage
6. Lubricate the final drive chain

**Monthly, or every 500 miles (800 km)**

1. Adjust the clutch
2. Check the engine idle speed
3. Adjust the throttle cable
4. Check the oil pump synchronisation and cable adjustment
5. Check, clean and re-gap the spark plug
6. Clean the front fork seal
7. Check the steering head bearing adjustment
8. Examine the wheels and check for bearing play
9. Adjust the brakes – drum brake
10. Adjust the final drive chain
11. Check the battery electrolyte level

**Six-monthly, or every 2500 miles (4000 km)**

1. Decarbonise the cylinder head, cylinder barrel and exhaust system
2. Examine and clean the air filter element
3. Renew the spark plug
4. Check the ignition timing
5. Lubricate all pivot points
6. Remove and lubricate all control cables
7. Lubricate the speedometer and tachometer cables
8. Check the degree of brake pad or shoe wear
9. Check the final drive chain and sprocket for wear

**Annually, or every 5000 miles (8000 km)**

1. Change the gearbox oil
2. Check the fuel system for contamination
3. Renew the air filter element
4. Renew the front fork oil
5. Lubricate the rear suspension pivot assemblies
6. Renew the disc brake hydraulic fluid and bleed the system
7. Remove, clean and lubricate the final drive chain
8. Check the battery specific gravity

**Two-yearly, or every 10 000 miles (16 000 km)**

1. Lubricate the steering head bearings
2. Lubricate the wheel bearings
3. Lubricate the speedometer drive gear
4. Lubricate the brake cam shafts – drum brakes
5. Renew the brake caliper piston seal and dust seal – disc brakes
6. Renew the master cylinder cup and dust seal – disc brakes

**Four yearly, or every 20 000 miles (32 000 km)**

1. Renew the fuel feed pipe
2. Renew the brake hose – disc brakes

## Adjustment data

| | |
|---|---|
| Spark plug type | NGK or ND |
| UK AE/AR50 | B6ES or W20ES-U |
| US AR50 | B8ES or W24ES-U |
| UK AE/AR80 A1 | BP8ES or W24EP-U |
| US AR80 | BP7ES or W22EP-U |
| UK AE80 B1, AR80 C1 | BPR8ES or W24EPR-U |
| Spark plug gap | 0.7 – 0.8 mm (0.027 – 0.031 in) |
| Idle speed | 1250 rpm |
| Tyre pressures | Pressures vary according to individual models and weights – see Chapter 5 |

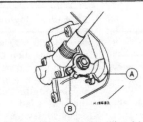

**Oil pump adjustment marks – UK models**
A  Pump pulley
B  Pump body index mark

**Oil pump adjustment marks – US models**
A  Pump body index mark
B  Pump pulley index mark
C  Idle mark (disregard)

## Recommended lubricants

| Component | Quantity | Type/Viscosity |
|---|---|---|
| 1 Engine: AE50 and 80 | 1.3 lit (2.3 Imp pt) | Good quality air-cooled 2-stroke engine oil |
| AR50 and 80 | 1.2 lit (2.1 Imp pt, 1.3 US qt) | |
| 2 Gearbox | 0.6 lit (1.06 Imp pt, 0.63 US qt) | SAE 10W/30 or 10W/40 SE or SF |
| 3 Front forks(per leg): AE50 and 80 | 92cc (3.24/3.11 Imp/US fl oz) | SAE 5W/20 |
| AR50 and 80 | 87cc (3.06/2.94 Imp/US fl oz) | |
| 4 Final drive chain | As required | Aerosol chain lubricant |
| 5 Wheel bearings | As required | High melting point grease |
| 6 Steering head bearings | As required | High melting point grease |
| 7 Rear suspension pivots | As required | High melting point grease |
| 8 Pivot points | As required | High melting point grease or gearbox oil |
| 9 Control cables | As required | Light machine oil |
| 10 Speedometer/ tachometer cables | As required | High melting point grease |
| 11 Air filter element | As required | SAE 30 |
| 12 Drum brake cam | As required | Brake grease |
| 13 Disc brake caliper spindles | As required | Poly Butyl Cuprysil grease |
| 14 Disc brake hydraulic system | As required | SAE J1703 (UK) or DOT 3 (US) |

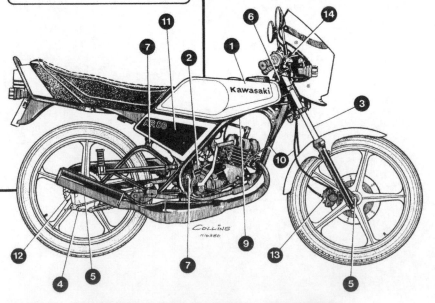

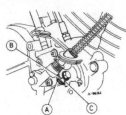

# ROUTINE MAINTENANCE GUIDE

*Refer to Chapter 7 for information relating to the 1984 on models*

# Routine maintenance

*Refer to Chapter 7 for information relating to the 1984 on models*

Periodic routine maintenance is a continuous process which should commence immediately the machine is used. The object is to maintain all adjustments and to diagnose and rectify minor defects before they develop into more extensive, and often more expensive problems.

It follows that if the machine is maintained properly, it will both run and perform with optimum efficiency, and be less prone to unexpected breakdowns. Regular inspection of the machine will show up any parts which are wearing, and with a little experience, it is possible to obtain the maximum life from any one component, renewing it when it becomes so worn that it is liable to fail.

Regular cleaning can be considered as important as mechanical maintenance. This will ensure that all the cycle parts are inspected regularly and are kept free from accumulations of road dirt and grime.

Cleaning is especially important during the winter months, despite its appearance of being a thankless task which very soon seems pointless. On the contrary, it is during these months that the paintwork, chromium plating, and the alloy casings suffer the ravages of abrasive grit, rain and road salt. A couple of hours spent weekly on cleaning the machine will maintain its appearance and value, and highlight small points, like chipped paint, before they become a serious problem.

The various maintenance tasks are described under their respective mileage and calendar headings, and are accompanied by diagrams and photographs where pertinent.

It should be noted that the intervals between each maintenance task serve only as a guide. As the machine gets older, or if it is used under particularly arduous conditions, it is advisable to reduce the period between each check.

Although no special tools are required for routine maintenance, a good selection of general workshop tools is essential. Included in the tools must be a range of metric ring or combination spanners, a selection of crosshead screwdrivers, and two pairs of circlip pliers, one

external opening and the other internal opening. Additionally, owing to the extreme tightness of most casing screws on Japanese machines, an impact screwdriver, together with a choice of large or small crosshead screw bits, is absolutely indispensable. This is particularly so if the engine has not been dismantled since leaving the factory.

**General note:**

The following maintenance schedule is based on that recommended by Kawasaki, with additions which the Author thinks necessary. Certain monthly tasks, such as wheel examination and spark plug check, should be approached with a degree of common sense and if thought necessary, carried out at a revised service interval. On no account ignore a maintenance task.

---

**Daily**

---

## 1   Engine oil level check

With the engine running and the machine in gear, check the oil level warning light is off. If the light remains on, stop the engine immediately and replenish the tank with a good quality non-diluent 2-stroke oil. The engine will be severely damaged if run with the light on. It is recommended that the level is maintained to within approximately an inch of the filler hole to allow good reserve. The filler cap is situated forward of the fuel tank.

## 2   Disc brake fluid level check

Whilst sat astride the machine, centre the handlebars and check that the brake fluid level is visible between the level marks of the reservoir sight glass. If the level is too low, top up the reservoir with SAE J1703 (UK) or DOT 3 (US) brake fluid before riding the machine. If the level has fallen rapidly, check the system for leaks.

Replenish the engine oil

Check the brake fluid level (AR models)

### 3  Stop lamp check

Before taking the machine on the road, check the stop lamp operates directly each brake is applied. Refer to Chapter 6 for stop lamp switch adjustment. The stop lamp and direction indicators give warning of your intentions to other drivers; a defective lamp can cause misunderstanding leading to an accident.

### Weekly, or every 125 miles (200 km)

### 1  Safety check

Give the machine a thorough inspection, checking for loose fasteners, frayed control cables, oil and petrol leaks, etc.

### 2  Legal check

Check the lights and horn are working properly and clean the lenses. It is an offence to ride a machine with defective lights, even in daylight. The horn in working order is also a statutory requirement.

### 3  Gearbox oil level check

Run the engine for a few minutes. Remove the filler plug just above the kickstart lever and wipe clean its dipstick. Lift the machine off its stand so that it is vertical and check that the oil level coincides with the centre hole of the plug dipstick with the plug screwed fully in. Remove or add oil as necessary.

### 4  Control cable lubrication

Apply a few drops of motor oil to the exposed inner of each cable. This will prevent the cables drying up before the more thorough 6 monthly lubrication.

### 5  Tyre pressure check

Check pressures with the tyres cold, using an accurate gauge. Purchase a pocket gauge to offset inaccuracies between garage forecourt instruments. Refer to Chapter 5 for correct pressures.

### 6  Tyre wear and damage check

Inspect the tyres for splits or cracks which may develop into serious faults. Remove small stones, etc from between the tread blocks. Renew a tyre with a tread depth less than 2 mm (0.08 in), or if it has any bald patches.

### 7  Final drive chain lubrication

With the chain fitted, spray it with one of the proprietary chain greases sold in aerosol form. Engine oil can be used but its effective life is limited due to the speed with which it is flung off the chain.

### Monthly, or every 500 miles (800 km)

Complete the preceding tasks and carry out the following:-

### 1  Clutch adjustment

Adjustment is correct when there is 2 – 3 mm (0.08 – 0.12 in) of free play in the cable inner, measured between the pivot end of the handlebar lever and its retaining clamp, If necessary, obtain this play by rotating the adjuster at the crankcase end of the cable, after having released its locknut. Retighten the locknut on completion.

### 2  Idle speed check

Start the engine and let it reach normal operating temperature. Allow the engine to idle. If the idle speed is not within 1200 – 1300 rpm, then adjust as follows. Note Section 7, Chapter 2 (US only). Set the throttle stop screw to give the slowest possible idle speed. On all models except the US AR80, turn the pilot screw in by a fraction of a turn at a time until the engine begins to falter; now turn it out whilst counting the number of turns required to reach the point where the engine again begins to falter. The correct pilot screw position is mid-way between the two extremes, where the engine is idling at its fastest. This should be close to the specified setting, see Chapter 2. If necessary, turn the throttle stop screw to achieve the specified idle speed. Where no tachometer is fitted, adjust to the lowest stable idle speed.

If the idle speed cannot be lowered, check throttle cable adjustment. If the speed varies with handlebar movement, check cable routing and condition.

Guard against the possibility of incorrect adjustment which will result in a weak mixture. Two-stroke engines are very susceptible to this type of fault, causing rapid overheating and often subsequent engine seizure. Changes in carburation leading to a weak mixture will occur if the air cleaner is removed or disconnected, or the exhaust system is tampered with.

### 3  Throttle cable adjustment

Adjustment is correct when there is 2 – 3 mm (0.08 – 0.12 in) of free play in the cable inner, measured by rotation of the twistgrip. Obtain this play by rotating the adjuster at the handlebar end of the cable, after having released its locknut. Retighten the locknut on completion. If adjustment is not available at the handlebar adjuster, move to the adjuster at the carburettor top.

Because the throttle and oil pump cables are interconnected and adjustment of one affects the other, proceed to the following paragraph.

Replenish the gearbox oil ...

... and check the oil level coincides with the centre hole of the dipstick

An aerosol lubricant should be used on the final drive chain

Check the clutch adjustment ...

... and if necessary, carry out clutch cable adjustment

Adjust the pilot screw (A) and throttle stop screw (B)

Check the throttle cable adjustment

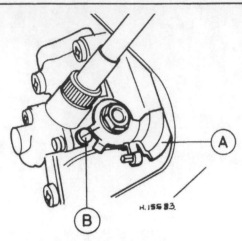

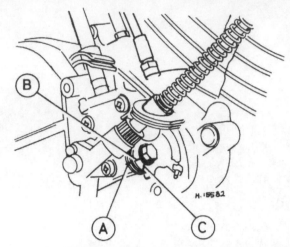

**Oil pump adjustment – UK models**

*A  Pulley*            *B  Pump body index mark*

**Oil pump adjustment – US models**

*A  Pump body index mark*        *C  idle mark (disregard)*
*B  Synchronisation mark*

## 4   Oil pump synchronisation and cable adjustment

Note the comments given in Chapter 7, Section 2 regarding the revised procedure.

### UK models

Start the engine and let it reach normal operating temperature. Allow the engine to idle and check the idle speed is correct. Stop the engine and check the throttle cable adjustment. Remove the pump cover and check that the cable is correctly routed over the pump pulley with its end secured by the retaining clip. Check the cable outer is correctly located in its adjuster.

Restart the engine and allow it to idle. Whilst listening to the exhaust note, slowly open the throttle twistgrip until engine speed just begins to move above idle. Hold the twistgrip in position and check the position of the pump pulley in relation to the pump body. If the marks on the pulley and body align, adjustment is correct.

### US models

Check throttle cable adjustment is correct. Remove the pump cover and check that the cable is correctly routed over the pump pulley with its end secured by the retaining clip. Check the cable outer is correctly located in its adjuster.

Ensure that the engine is switched off. Hold the throttle twistgrip fully open and check the position of the pump pulley in relation to the pump body. Adjustment is correct if the synchronisation mark is in exact elignment with the mark on the body – see accompanying figure.

### All models

If necessary, carry out adjustment by releasing the locknut of the cable adjuster and turning the adjuster until the marks align. Take care to keep the twistgrip held in position during adjustment. Retighten the adjuster locknut on completion. Check the twistgrip for free play and refit the pump cover.

## 5   Spark plug check

Check the plug type, see Chapter 3, and replace if incorrect. Clean the electrodes with a small brass-wire brush, removing stubborn carbon deposits by scraping with a pocket knife whilst taking care not to chip the porcelain insulator of the centre electrode. Pass a small fine file or emery paper between the electrode faces and remove all traces of abrasive material from the plug on completion of cleaning.

Some local garages and dealers have plug cleaning machines and will clean plugs for a nominal fee. This is a very efficient method but check there is no blasting medium rammed between the insulator and plug body before fitting.

Reset the electrode gap to 0.7 – 0.8 mm (0.027 – 0.031 in), using a feeler gauge which should be a light sliding fit. Bend the outer electrode to alter the gap; never apply force to the centre electrode.

Lightly smear the plug threads with graphite grease. Check the sealing washer is fitted. Do not overtighten the plug, see torque settings, Chapter 1. Always carry a spare plug.

Before reconnecting the suppressor cap, check its seals and renew if damaged or perished. Check the cap is a good firm fit; on early models it contains the suppressor which eliminates TV and radio interference.

Check the oil pump adjuster locknut is tight

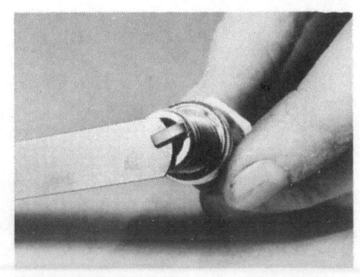

Check the spark plug electrode gap

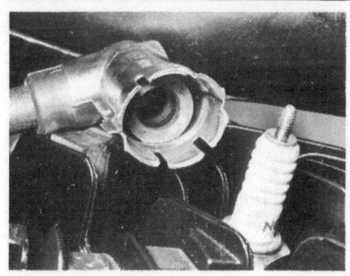
Examine the suppressor cap seals

### 6 Front fork seal clean

Disregard this operation if gaiters have been fitted in place of the standard dust excluders.

Wipe the fork stanchion clean and pull the dust excluder off the lower leg. Using a clean, soft rag, carefully wipe any dirt or moisture off the oil seal and the inside of the dust excluder. A leaking oil seal or split dust excluder must be renewed, see Chapter 4. Refit the excluder over the lower leg, checking it is a good firm fit.

### 7 Steering head bearing check and adjustment

Position a stout wooden crate or blocks beneath the engine so that the machine is well supported, with the front wheel clear of the ground. Grasp the fork legs near the wheel spindle and push and pull firmly in a fore and aft direction. If play is evident between the fork yokes and steering head, bearing adjustment is necessary. Play will cause fork judder or imprecise handling. If necessary, adjust as follows.

On AE models, remove the fuel tank front cover whilst noting the fitted position of the spacers and washers retained by its two securing screws. Protect the tank with a piece of blanket. Detach the handlebars from the fork upper yoke and ease them clear of the steering head. Loosen the upper yoke clamp bolts.

On AR models, refer to Chapter 2 and remove the fuel tank. Loosen the fork lower yoke clamp bolts.

On all models, release the steering stem top bolt. Using a C-spanner, back off the bearing adjuster ring one or two turns only. Retighten the adjuster ring to 2.0 kgf m (14.5 lbf ft) torque loading. If no torque wrench is available, then tighten the ring until slight resistance is felt and then continue to turn for another $\frac{1}{16}$ turn. Retighten the top bolt to 2.0 kgf m (14.5 lbf ft) torque loading.

Note that it is possible to place a pressure of several tons on the bearings by overtightening, even though the handlebars may turn quite freely. Overtight bearings will cause the machine to roll at low speeds and give imprecise steering. Adjustment is correct if there is no play and the handlebars swing freely to full lock in each direction. A light tap on each end of the handlebars should cause them to swing.

With adjustment complete, retighten the fork yoke clamp bolts to the specified torque loading and refit all disturbed components.

### 8 Wheel examination and bearing check

Raise clear of the ground the wheel to be examined, using blocks positioned beneath the engine. Check the wheel spins freely; if necessary, slacken the brake adjuster and detach the final drive chain.

**AE models**

Examine the rim for serious corrosion or impact damage. Slight deformities can often be corrected by adjusting spoke tension. Serious damage and corrosion will necessitate renewal, which is best left to an expert. A light alloy rim will prove more corrosion resistant.

Place a wire pointer close to the rim and rotate the wheel to check it for runout. If the rim is more than 2.0 mm (0.08 in) out of true in the radial or axial planes, check spoke tension by tapping them with a screwdriver. A loose spoke will sound quite different to those around it. Worn bearings will also cause rim runout.

Adjust spoke tension by turning the square-headed nipples with the appropriate spoke key which can be purchased from a dealer. With the spokes evenly tensioned, remaining distortion can be pulled out by tightening the spokes on one side of the wheel and slackening those directly opposite. This will pull the rim across whilst maintaining spoke tension.

More than slight adjustment will cause the spoke ends to protrude through the nipple and chafe the inner tube, causing a puncture. Remove the tyre and tube and file off the protruding ends. The rim band protects the tube against chafing, check it is in good condition before fitting.

Check spoke tension and general wheel condition regularly. Frequent cleaning will help prevent corrosion. Replace a broken spoke immediately because the load taken by it will be transferred to adjacent spokes which may fail in turn.

**AR models**

Carefully check the complete wheel for cracks and chipping, particularly at the spoke roots and the edge of the rim. As a general rule a damaged wheel must be renewed, as cracks will cause stress points which may lead to sudden failure under heavy load. Small nicks may be radiused carefully with a fine file and emery paper (No 600 – No 1000) to relieve the stress. If in doubt seek advice from a Kawasaki dealer.

Place a wire pointer close to the rim and rotate the wheel to check it for runout. If the rim is more than 0.5 mm (0.02 in) out of true in the axial plane or more than 0.8 mm (0.03 in) out of true in the radial plane, the manufacturer recommends that the wheel be renewed. This is, however, a question of perfection; a run out somewhat greater than this can probably be accommodated without noticeable effect on steering. No means is available for straightening a warped wheel without resorting to the expense of having the wheel skimmed on all faces. If warpage was caused by impact during an accident, the safest measure is to renew the wheel complete. Worn wheel bearings may cause rim run out.

Each wheel is covered with a coating of lacquer, to prevent corrosion. If damage occurs to the wheel and the lacquer finish is penetrated, the bared aluminium alloy will soon start to corrode. A whitish grey oxide will form over the damaged area, which in itself is a protective coating. This deposit however, should be removed carefully as soon as possible and a new protective coating of lacquer applied.

**All models**

An out of balance wheel will produce a hammering effect through the steering at high speed. Spin the wheel several times. A well balanced wheel will come to rest in any position. One that comes to rest in the same position will have its heaviest part downward and weights must be added to a point exactly opposite until balance is achieved. Where the tyre has a balance mark on its sidewall (usually a coloured spot), check it is in line with the valve.

Grasp the end of the wheel spindle and spin the wheel. Excessive vibrations felt through the spindle will indicate bearing wear as will a rumble emitted from the wheel centre. Renew the worn bearing(s) as soon as possible, see Chapter 5.

### 9 Brake adjustment

This check applies only to drum brakes. Automatic compensation for wear of disc brake components obviates the need for frequent adjustment checks.

**Front**

Adjustment is correct when there is 4 – 5 mm (0.16 – 0.20 in) of free play in the cable inner, measured between the pivot end of the handlebar lever and its retaining clamp. Rotate the cable adjuster at the wheel end to obtain this play.

**Rear**

Adjustment is correct when there is 25 mm (1.0 in) of movement at the brake pedal footplate. Rotate the adjuster at the wheel end of the brake rod to obtain this movement. Reset the stop lamp switch if necessary, see Chapter 6.

**Both**

On completion, check the brake for correct operation. Spin the wheel and apply the brake; there should be no indication of the brake binding. Back off the adjuster until binding disappears and recheck operation.

### 10 Final drive chain adjustment

Check chain adjustment with the machine unloaded and resting on its prop stand. Rotate the rear wheel and find the chain tight spot, positioning it in the centre of the lower chain run. The correct up and down movement is 30 – 35 mm (1.2 – 1.4 in) measured at the tight spot.

To adjust, loosen the wheel spindle and torque arm to brake backplate nuts. Release both adjuster locknuts and tighten the chain by turning each adjuster nut clockwise an equal number of turns. Verify wheel alignment by checking each adjuster mark is aligned with the same mark on each fork end.

With adjustment correct, retighten the adjuster locknuts. Lightly tighten the spindle nut, spin the wheel and forcefully apply the rear brake; this will centre the brake backplate in the drum. Tighten the spindle nut to the specified torque loading, see Chapter 5, recheck chain adjustment and torque load the torque arm nut, checking its spring-pin is correctly fitted.

Check the wheel spins freely. If necessary, adjust brake and stop lamp operation.

An overtight chain will place excessive loads on the gearbox and rear wheel bearings, leading to their early failure. It will also absorb a surprising amount of power.

### 11 Battery electrolyte level check

Check the electrolyte level is between the upper and lower marks on the battery case. If necessary, replenish each cell with distilled water. Do not overfill. Check the vent pipe is attached and correctly routed.

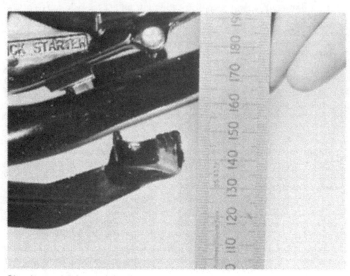

Check rear brake pedal adjustment

Check final drive chain adjustment

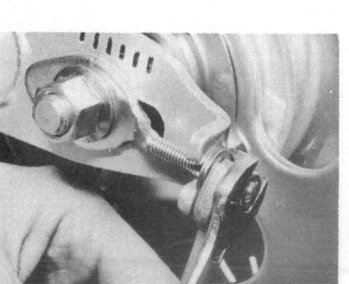

Retighten the chain adjuster locknuts

Check the battery electrolyte level ...

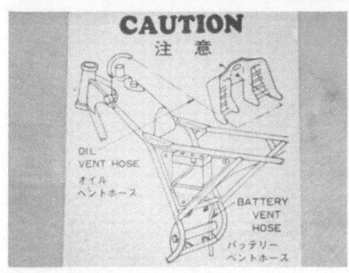

... and note the instructions on vent pipe routing

## Six monthly, or every 2500 miles (4000 km)

Complete the preceding tasks and carry out the following:-

### 1 Decarbonisation

**Cylinder head and barrel**

Refer to Chapter 1 and remove the cylinder head and barrel. It is necessary to remove all carbon from the head, barrel and piston crown whilst avoiding removal of the metal surfaces on which it is deposited. Take care when dealing with the soft alloy head and piston. Never use a steel scraper or screwdriver. A hardwood, brass or aluminium scraper is ideal as these are harder than the carbon but no harder than the underlying metal. With the bulk of carbon removed, use a brass wire brush. Finish the head and piston with metal polish; a polished surface will slow the subsequent build-up of carbon. Clean out the barrel ports to prevent the restriction of gas flow. Remove all debris by washing each component in paraffin whilst observing the necessary fire precautions.

**Exhaust system**

Remove the exhaust silencer baffle by removing its retaining screw(s) with washer(s), gripping its end with pliers or a mole wrench and pulling it from position. If the baffle has seized in position, pass a tommy bar through the mole wrench and strike it rearwards with a hammer.

Remove and discard any asbestos tape wrapped around the baffle; it need not be replaced. If the build up of carbon and oil on the baffle is not too great, wash it clean with petrol whilst taking the necessary fire precautions. Heavy deposits on the baffle may well indicate similar contamination within the silencer and pipe in which case the system should be removed, see Chapter 2. Clean the baffle by running a blowlamp along its length to burn off the deposits, waiting for it to cool and tapping it sharply with a length of hardwood to dislodge any remaining deposits. Finish with a wire brush and check the baffle holes are clear.

Suspend the system from its rearmost end(s). Block the lower end(s) with a cork or wooden bung. On AE models, remove the expansion chamber to silencer seal. Mix up a caustic soda solution (3 lb to over a gallon of fresh water), adding the soda to the water gradually, whilst stirring. Do not pour water into a container of soda, this will cause a violent reaction to take place. Wear proper eye and skin protection; caustic soda is very dangerous. Eyes and skin contaminated by soda must be immediately flushed with fresh water and examined by a doctor. The solution will react violently with aluminium alloy, causing severe damage to any components.

Fill the system with solution, leaving the upper end(s) open. Leave the solution overnight to allow its dissolving action to take place. Ventilate the area to prevent the build-up of noxious fumes. On completion, carefully pour out the solution and flush the system through with clean, fresh water.

Refit the system. Lightly smear the baffle mating surface and retaining screw threads to prevent seizure. Renew the screw spring washer(s) if flattened.

Do not modify the baffle or run the machine with it removed. This will result in less performance and affect the carburation.

### 2 Air filter element examination and cleaning

Detach the right-hand sidepanel. Remove the element cover (3 screws) and the element. Renew the element if hardened or badly clogged.

If serviceable, immerse the element in white spirit (Stoddard solvent in the US), gently squeezing it to remove all oil and dust. Remove excess solvent by wrapping the element in clean rag and pressing it between the palms of the hands; wringing it out will cause damage. Allow a short time for any remaining solvent to evaporate. Reimpregnate the element with clean SAE 30 oil and squeeze as dry as possible.

Clean the element housing to cover mating surfaces. When fitting, position the element and its cover correctly. Air which bypasses the element will carry dirt into the carburettor and crankcase and will weaken the fuel/air mixture.

If riding in a particularly dusty or moist atmosphere, increase the frequency of cleaning the element. Never run the engine without the element fitted; the carburettor is jetted to compensate for its being fitted and the resulting weak mixture will cause overheating of the engine.

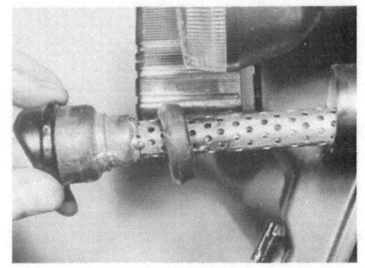

Withdraw the exhaust silencer baffle (AE shown)

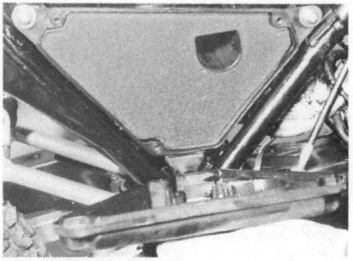

Expose the air filter element

### 3 Spark plug renewal

Remove and discard the existing plug, regardless of condition. It will have passed peak efficiency. Fit a new plug of the correct type, see Chapter 3. Before fitting, check the electrode gap is 0.7 – 0.8 mm (0.027 – 0.031 in), smear the plug threads sparingly with graphite grease and check the aluminium crush washer is fitted.

### 4 Ignition timing check

Even a small error in timing will reduce the engine performance, possibly causing damage as a result of overheating.

Remove the flywheel generator cover. Degrease the timing marks on the rotor wall and crankcase and coat each one with a trace of white paint. This should only be necessary if the light from the stroboscopic lamp being used is weak or conditions are bright. Connect the 'strobe' to the HT lead, following the maker's instructions.

Start the engine and aim the 'strobe' at the timing marks. The light will 'freeze' the moving mark on the rotor in one position; this mark must be directly in line with that on the crankcase with the engine running at 3000 rpm.

Timing can be adjusted by removing the rotor (see Chapter 1), loosening the stator retaining screws, rotating the stator a small amount in the required direction (clockwise to advance timing), retightening the screws, fitting the rotor back onto the crankshaft and rechecking the timing. Repeat this operation until the timing is correct, check tighten the stator retaining screws, refit the rotor, remove all test equipment and refit the spark plug and rotor cover.

### 5 General lubrication

Work around the machine, applying grease or oil to any pivot points. These should include the footrests, prop stand, brake pedal, twistgrip, brake cable trunnions, handlebar levers and kickstart.

### 6 Control cable lubrication

Do not lubricate nylon lined cables which may have been fitted as replacements; this can cause total seizure by swelling the nylon.

Lubricate each cable with motor oil using either a hydraulic oiler, which can be purchased from a dealer, or by using the method indicated in the accompanying figure.

### 7 Speedometer and tachometer cable lubrication

Remove the cable and withdraw its inner. If this is not possible, a badly seized cable will have to be renewed. Wipe the inner with a petrol soaked rag and examine it for broken strands or damage. Do not pull the inner through the hand, broken strands will tear the skin; use a piece of rag.

Lubricate the inner with high melting-point grease. Do not grease the six inches closest to the instrument head; grease will work into the instrument and immobilise the sensitive movement.

### 8 Brake wear check

**Drum**

An indicator line on the brake backplate and a pointer on the cam spindle provide an indication of shoe lining wear. The pointer must be inside the arc of the indicator line with the brake fully applied. If not, renew the shoes.

**Disc**

Remove the brake caliper to fork leg mounting bolts and ease the caliper clear of the disc. Take care not to squeeze the brake lever as this will expel the piston and cause fluid loss. At no time breathe in brake dust, it contains asbestos and is harmful to health. Do not allow the caliper to hang from the brake hose.

Note the fitted position of both brake pads and the anti-rattle spring before removing them from the caliper. Renew the spring if cracked, distorted or broken. Wipe clean the pads and caliper with a soft dry rag.

Measure the lining thickness of each pad. Renew both pads as a set if either thickness is less than 1.0 mm (0.04 in). Check the piston head has remained in position and fit the serviceable pads and spring in the caliper. When fitting new pads, push the piston fully into the caliper to create sufficient clearance between the pads to allow fitting over the disc.

Carefully slide the caliper over the disc, avoiding damage to the pads. Fit the caliper mounting bolts and torque load them to 3.3 kgf m (24.0 lbf ft). Before taking the machine on the road, pump the brake lever several times to restore full braking power and check for correct operation.

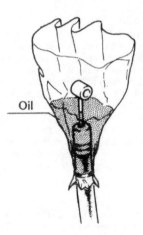

Control cable lubrication

The ignition timing marks must be directly in line

Drum brakes incorporate a wear indicator

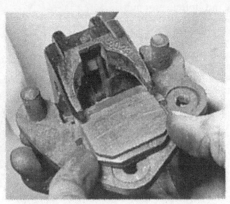

Check the disc brake caliper piston head is correctly located ...

... fit the first brake pad over the piston ...

... and fit the second brake pad

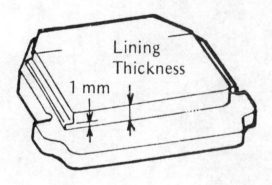

**Brake pad lining wear limit**

### 9 Final drive chain and sprocket wear check

Renew either sprocket if its teeth are hooked or badly worn. It is bad practice to renew one sprocket on its own; both drive sprockets should be renewed, preferably with the chain. Running old and new parts together will result in rapid wear.

To renew the wheel sprocket, remove the wheel and bend back the locking tab from each sprocket securing nut or bolt. Remove the nuts or bolts, tab washers and sprocket. When fitting, do not rebend the locking tabs, renew the washers if necessary. Tighten the nuts or bolts evenly and in a diagonal sequence to a torque loading of 2.1 kgf m (15.0 lbf ft).

To renew the gearbox sprocket, remove its cover, detach the chain and remove the sprocket retaining circlip. Renew the circlip if cracked or distorted. With the new sprocket fitted, check the circlip is a firm fit on the shaft.

Check the chain for obvious damage, such as cracked or missing rollers, and renew if necessary. To determine wear, either hang a 10 kg (22 lb) weight from the centre of its lower run or wind the adjusters fully in to stretch the chain. Select a length of chain in the middle of the upper run and count off 21 pins, that is a 20 pitch length. Measure the distance between the 1st and 21st pins. If the distance exceeds 259 mm (10.2 in), renew the chain. Because the chain will wear unevenly, take this measurement at several points along its length.

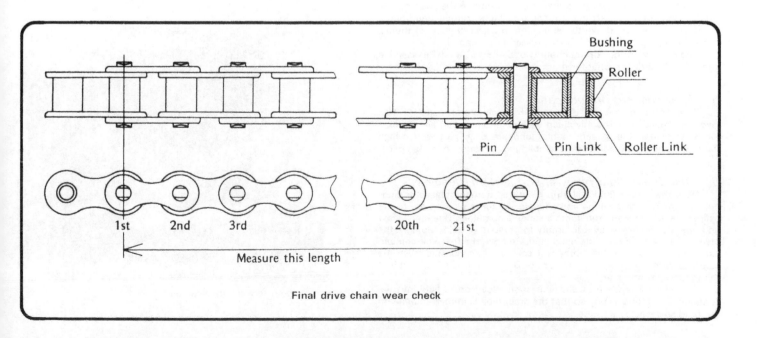

**Final drive chain wear check**

To renew the chain, loosen the wheel spindle nut, torque arm nut and adjuster nuts before pushing the wheel forward. Rotate the wheel so the split link is at the wheel sprocket. Using flat-nose pliers, remove the spring clip from the link and withdraw the link. Join the new and old chains and use the old chain to pull the new one over the gearbox sprocket. Connect the ends of the new chain with the new link provided, the spring clip must have the side plate fitted beneath it, be seated correctly and have its closed end facing the direction of chain travel. Readjust the chain; see monthly maintenance.

Replacement chains are available from Renold Limited, the British chain manufacturer. When ordering, quote chain size, the number of links and machine type.

The chain split link must be correctly fitted

### Annually, or every 5000 miles (8000 km)

Complete the preceding tasks and carry out the following:-

### 1  Gearbox oil change

Position a container of sufficient capacity beneath the engine. Run the engine to warm and thin the oil. Remove the gearbox oil drain plug and sealing washer from the underside of the crankcase. Whilst waiting for the oil to drain, examine the sealing washer and renew, if damaged.

On completion of draining, refit the plug and tighten to 2.0 kgf m (14.5 lbf ft). Remove the filler plug and wipe clean its dipstick. Replenish the gearbox with 0.6 lit (1.06 Imp pt, 0.63 US qt) of SAE 10W/30 or 10W/40 SE or SF oil. Wait a few minutes before lifting the machine off its stand so that it is vertical and checking that the oil level coincides with the centre hole of the filler plug dipstick with the plug screwed fully in. Remove or add oil as necessary.

### 2  Fuel system check

Kawasaki recommend that the fuel system be regularly checked for contamination. The first signs of contamination will be found at the tap filter stack and carburettor float chamber. Contamination can cause fuel starvation, see Fault diagnosis. If necessary, dismantle and clean each system component, see Chapter 2.

### 3  Air filter element renewal

Kawasaki recommend that the air filter element be renewed at every 5000 mile service interval.

### 4  Front fork oil renewal

Remove each fork leg, see Chapter 4. Remove the leg top bolt or plug and remove the spring retaining plug. Remove the spacer tube, washer and spring. Invert the leg over a suitable container and pump the lower leg up and down the stanchion to assist draining of the oil. Renew the plug or bolt O-ring if flattened or damaged.

Replenish with the correct quantity of SAE 5W/20 oil. Reassemble and refit the leg, see Chapter 4.

### 5  Rear suspension lubrication

Refer to Chapter 4 and dismantle the swinging arm and Uni-trak pivot assemblies. Thoroughly clean each component part and check it for wear or damage. If serviceable, lubricate each part with a high melting point grease before reassembling. Renew unserviceable parts.

### 6  Disc brake fluid renewal and system bleeding

Note that brake fluid will strip paint and damage plastic components, wipe up any spillage immediately. Attach a plastic tube to the caliper bleed screw and run it into a suitable container. Open the screw and operate the brake lever to empty the system of fluid. Nip tight the bleed screw and discard the used fluid. Remove the reservoir cap and diaphragm. Renew the diaphragm if split or perished. Replenish the reservoir with clean SAE J1703 (UK) or DOT (US) brake fluid and refit the diaphragm and cap.

To bleed the system of air, pour some clean brake fluid into the container used for draining so that the drain tube is immersed beneath the fluid surface.

Open the disc brake caliper bleed screw to release fluid ...

... replenish the reservoir with clean fluid ...

**Electrode gap check** - use a wire type gauge for best results

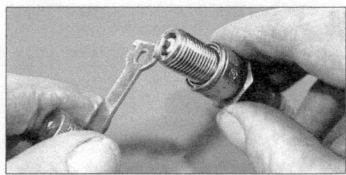

**Electrode gap adjustment** - bend the side electrode using the correct tool

**Normal condition** - A brown, tan or grey firing end indicates that the engine is in good condition and that the plug type is correct

**Ash deposits** - Light brown deposits encrusted on the electrodes and insulator, leading to misfire and hesitation. Caused by excessive amounts of oil in the combustion chamber or poor quality fuel/oil

**Carbon fouling** - Dry, black sooty deposits leading to misfire and weak spark. Caused by an over-rich fuel/air mixture, faulty choke operation or blocked air filter

**Oil fouling** - Wet oily deposits leading to misfire and weak spark. Caused by oil leakage past piston rings or valve guides (4-stroke engine), or excess lubricant (2-stroke engine)

**Overheating** - A blistered white insulator and glazed electrodes. Caused by ignition system fault, incorrect fuel, or cooling system fault

**Worn plug** - Worn electrodes will cause poor starting in damp or cold weather and will also waste fuel

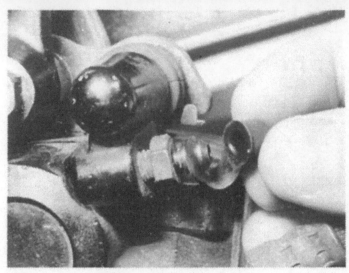

... always refit the bleed screw cover

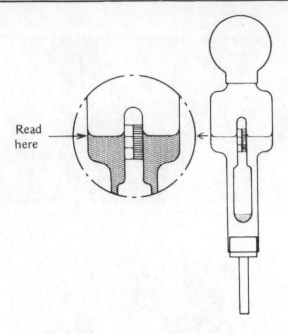

Battery specific gravity check hydrometer reading

Pump the brake lever several times in rapid succession, finally holding the lever in the 'fully on' position. Loosen the bleed screw one quarter of a turn so that the fluid runs down the tube into the container. This will cause the pressure within the system to be released thereby causing the brake lever to move from its 'fully on' position to touch the throttle twistgrip. Directly this happens, nip the bleed screw shut. As the fluid is ejected from the bleed screw, the level in the reservoir will fall. Take care that the level does not drop too low whilst the operation continues, otherwise air will re-enter the system, necessitating a fresh start.

Repeat the above procedure until no further air bubbles emerge from the tube end. Hold the brake lever against the twistgrip and tighten the bleed screw. Remove the tube **after** the bleed screw is closed.

Check the brake action for sponginess, which usually denotes there is still air in the system. If the action is spongy, continue the bleeding operation in the same manner, until all traces of air are removed.

Check the fluid level in the reservoir. Check the system for leaks with the brake fully applied. Recheck brake action and check the stop lamp for correct operation. Do not reuse the fluid in the container, it will be aerated and may have absorbed moisture. Refit the bleed screw cover.

### 7　Final drive chain clean and lubrication

Position a stout wooden crate or blocks beneath the engine so that the machine is well supported with the rear wheel clear of the ground. Lay a length of clean rag beneath the chain. Rotate the wheel to align the chain split link with the wheel sprocket. Using flat-nose pliers, remove the spring clip from the link and withdraw the link.

If an old chain is available, connect it to the one on the machine before it is run off the gearbox sprocket then it can be used to pull the greased chain back over the sprocket. Otherwise the gearbox sprocket cover will have to be removed.

Wash the chain thoroughly in petrol, observing the necessary fire precautions. Dry the chain and immerse it in a molten lubricant such as Linklyfe or Chainguard. Follow the manufacturer's instructions.

Refit the chain. The spring clip of the link must have the side plate fitted beneath it, be seated correctly and have its closed end facing the direction of chain travel. Check chain adjustment.

### 8　Battery specific gravity check

Remove the battery and check the specific gravity of its acid is 1.260 at 20°C (68°F). Take the reading at the top of the meniscus with the hydrometer vertical.

Protect the eyes and skin against accidental spillage of acid. Eyes contaminated with acid must be immediately flushed with copious amounts of fresh water and examined by a doctor, likewise the skin.

### Two yearly, or every 10 000 miles (16 000 km)

Complete the preceding tasks and carry out the following:

### 1　Steering head bearing lubrication

The relevant information for carrying out this task is contained in Chapter 4.

### 2　Wheel bearing lubrication

With each wheel removed, its bearings can be removed for the purposes of examination and lubrication; see Chapter 5.

### 3　Speedometer drive gear lubrication

The relevant information for carrying out this task is contained in Chapter 5.

### 4　Drum brake cam shaft lubrication

The relevant information for carrying out this task is contained in Chapter 5, Section 10. Failure to lubricate a shaft could well result in its seizing during operation of the brake with disastrous consequences.

### 5　Disc brake caliper piston seal and dust seal renewal

In the interests of safety, Kawasaki recommend that these seals are renewed as a matter of course. The relevant information for carrying out this task is contained in Chapter 5.

### 6　Disc brake master cylinder cup and dust seal renewal

To obviate the risk of brake failure, renew these seals as a matter of course. Refer to Chapter 5 for the relevant information.

### Four yearly, or every 20 000 miles (32 000 km)

Complete the preceding tasks and carry out the following:-

### 1　Fuel feed pipe renewal

To anticipate any risk of fuel leakage from a hardened, split or perished pipe, renew the pipe as a matter of course.

### 2　Disc brake hose renewal

To obviate any risk of brake failure, renew the hydraulic hose as a matter of course. Refer to Chapter 5 for the relevant information.

**Additional maintenance items**

## Cleaning the machine

A machine cleaned regularly will corrode less readily and hence maintain its market value. It will also be more approachable when the time comes for maintenance. Loose or failing components are more easily spotted when not obscured by dirt and oil.

## Metal components

Use a sponge and copious amounts of warm soapy water to wash surface dirt from these components. Remove oil and grease with a solvent such as 'Gunk' or 'Jizer', working it in with a stiff brush when the component is still dry and rinsing off with fresh water. Keep water out of the carburettor, air filter and electrics.

Except where an aluminium alloy component is lacquered, use a polish such as Solvol Autosol to restore its original lustre. Apply wax polish to the painted components and a good chrome cleaner to those which are chromed. Keep the chain and control cables well lubricated to prevent the ingress of water and wipe the machine down if used in the wet.

## Plastic components

Do not use strong detergents, scouring powders or any abrasive when cleaning these components; anything but a mild solution of soapy water may well bleach or score the surface. On completion of cleaning wipe the component dry with a chamois leather. If the surface finish has faded, use a fine aerosol polish to restore its shine.

# Conversion factors

## Length (distance)

| | | | | | |
|---|---|---|---|---|---|
| Inches (in) | X 25.4 | = Millimetres (mm) | X 0.0394 | = Inches (in) |
| Feet (ft) | X 0.305 | = Metres (m) | X 3.281 | = Feet (ft) |
| Miles | X 1.609 | = Kilometres (km) | X 0.621 | = Miles |

## Volume (capacity)

| | | | | |
|---|---|---|---|---|
| Cubic inches (cu in; in³) | X 16.387 | = Cubic centimetres (cc; cm³) | X 0.061 | = Cubic inches (cu in; in³) |
| Imperial pints (Imp pt) | X 0.568 | = Litres (l) | X 1.76 | = Imperial pints (Imp pt) |
| Imperial quarts (Imp qt) | X 1.137 | = Litres (l) | X 0.88 | = Imperial quarts (Imp qt) |
| Imperial quarts (Imp qt) | X 1.201 | = US quarts (US qt) | X 0.833 | = Imperial quarts (Imp qt) |
| US quarts (US qt) | X 0.946 | = Litres (l) | X 1.057 | = US quarts (US qt) |
| Imperial gallons (Imp gal) | X 4.546 | = Litres (l) | X 0.22 | = Imperial gallons (Imp gal) |
| Imperial gallons (Imp gal) | X 1.201 | = US gallons (US gal) | X 0.833 | = Imperial gallons (Imp gal) |
| US gallons (US gal) | X 3.785 | = Litres (l) | X 0.264 | = US gallons (US gal) |

## Mass (weight)

| | | | | |
|---|---|---|---|---|
| Ounces (oz) | X 28.35 | = Grams (g) | X 0.035 | = Ounces (oz) |
| Pounds (lb) | X 0.454 | = Kilograms (kg) | X 2.205 | = Pounds (lb) |

## Force

| | | | | |
|---|---|---|---|---|
| Ounces-force (ozf; oz) | X 0.278 | = Newtons (N) | X 3.6 | = Ounces-force (ozf; oz) |
| Pounds-force (lbf; lb) | X 4.448 | = Newtons (N) | X 0.225 | = Pounds-force (lbf; lb) |
| Newtons (N) | X 0.1 | = Kilograms-force (kgf; kg) | X 9.81 | = Newtons (N) |

## Pressure

| | | | | |
|---|---|---|---|---|
| Pounds-force per square inch (psi; lbf/in²; lb/in²) | X 0.070 | = Kilograms-force per square centimetre (kgf/cm²; kg/cm²) | X 14.223 | = Pounds-force per square inch (psi; lbf/in²; lb/in²) |
| Pounds-force per square inch (psi; lbf/in²; lb/in²) | X 0.068 | = Atmospheres (atm) | X 14.696 | = Pounds-force per square inch (psi; lbf/in²; lb/in²) |
| Pounds-force per square inch (psi; lbf/in²; lb/in²) | X 0.069 | = Bars | X 14.5 | = Pounds-force per square inch (psi; lbf/in²; lb/in²) |
| Pounds-force per square inch (psi; lbf/in²; lb/in²) | X 6.895 | = Kilopascals (kPa) | X 0.145 | = Pounds-force per square inch (psi; lbf/in²; lb/in²) |
| Kilopascals (kPa) | X 0.01 | = Kilograms-force per square centimetre (kgf/cm²; kg/cm²) | X 98.1 | = Kilopascals (kPa) |
| Millibar (mbar) | X 100 | = Pascals (Pa) | X 0.01 | = Millibar (mbar) |
| Millibar (mbar) | X 0.0145 | = Pounds-force per square inch (psi; lbf/in²; lb/in²) | X 68.947 | = Millibar (mbar) |
| Millibar (mbar) | X 0.75 | = Millimetres of mercury (mmHg) | X 1.333 | = Millibar (mbar) |
| Millibar (mbar) | X 0.401 | = Inches of water (inH₂O) | X 2.491 | = Millibar (mbar) |
| Millimetres of mercury (mmHg) | X 0.535 | = Inches of water (inH₂O) | X 1.868 | = Millimetres of mercury (mmHg) |
| Inches of water (inH₂O) | X 0.036 | = Pounds-force per square inch (psi; lbf/in²; lb/in²) | X 27.68 | = Inches of water (inH₂O) |

## Torque (moment of force)

| | | | | |
|---|---|---|---|---|
| Pounds-force inches (lbf in; lb in) | X 1.152 | = Kilograms-force centimetre (kgf cm; kg cm) | X 0.868 | = Pounds-force inches (lbf in; lb in) |
| Pounds-force inches (lbf in; lb in) | X 0.113 | = Newton metres (Nm) | X 8.85 | = Pounds-force inches (lbf in; lb in) |
| Pounds-force inches (lbf in; lb in) | X 0.083 | = Pounds-force feet (lbf ft; lb ft) | X 12 | = Pounds-force inches (lbf in; lb in) |
| Pounds-force feet (lbf ft; lb ft) | X 0.138 | = Kilograms-force metres (kgf m; kg m) | X 7.233 | = Pounds-force feet (lbf ft; lb ft) |
| Pounds-force feet (lbf ft; lb ft) | X 1.356 | = Newton metres (Nm) | X 0.738 | = Pounds-force feet (lbf ft; lb ft) |
| Newton metres (Nm) | X 0.102 | = Kilograms-force metres (kgf m; kg m) | X 9.804 | = Newton metres (Nm) |

## Power

| | | | | |
|---|---|---|---|---|
| Horsepower (hp) | X 745.7 | = Watts (W) | X 0.0013 | = Horsepower (hp) |

## Velocity (speed)

| | | | | |
|---|---|---|---|---|
| Miles per hour (miles/hr; mph) | X 1.609 | = Kilometres per hour (km/hr; kph) | X 0.621 | = Miles per hour (miles/hr; mph) |

## Fuel consumption*

| | | | | |
|---|---|---|---|---|
| Miles per gallon, Imperial (mpg) | X 0.354 | = Kilometres per litre (km/l) | X 2.825 | = Miles per gallon, Imperial (mpg) |
| Miles per gallon, US (mpg) | X 0.425 | = Kilometres per litre (km/l) | X 2.352 | = Miles per gallon, US (mpg) |

## Temperature

Degrees Fahrenheit = (°C x 1.8) + 32

Degrees Celsius (Degrees Centigrade; °C) = (°F - 32) x 0.56

*It is common practice to convert from miles per gallon (mpg) to litres/100 kilometres (l/100km), where mpg (Imperial) x l/100 km = 282 and mpg (US) x l/100 km = 235

# Chapter 1  Engine, clutch and gearbox

*Refer to Chapter 7 for information relating to the 1984 on models*

## Contents

## Specifications

**Note**: The following specifications are for UK models only. Owners of US models check with the information at the end of this Section.

| Engine | AE and AR50 | AE and AR80 |
| --- | --- | --- |
| Type | Air cooled, single cylinder, two-stroke | |
| Bore | 39.0 mm (1.535 in) | 49.0 mm (1.929 in) |
| Stroke | 41.6 mm (1.638 in) | 41.6 mm (1.638 in) |
| Capacity | 49 cc (2.99 cu in) | 78 cc (4.76 cu in) |
| Compression ratio | 7.0:1 | 7.8:1 |

| Cylinder barrel | | |
| --- | --- | --- |
| Standard bore | 39.0 – 39.015 mm (1.535 – 1.536 in) | 49.0 – 49.015 mm (1.929 – 1.930 in) |
| Service limit | 39.10 mm (1.539 in) | 49.10 mm (1.933 in) |

| Cylinder head | | |
| --- | --- | --- |
| Distortion limit | 0.1 mm (0.004 in) | 0.1 mm (0.004 in)* |

*For AE80 B1 and AR80 A1 models see US entry*

| Piston | | |
| --- | --- | --- |
| Outside diameter | 38.070 – 38.985 mm (1.534 – 1.535 in) | 48.965 – 48.980 mm (1.927 – 1.928 in) |
| Service limit | 38.83 mm (1.529 in) | 48.82 mm (1.922 in) |
| Piston to bore clearance | 0.020 – 0.030 mm (0.0008 – 0.0012 in) | 0.030 – 0.040 mm (0.0012 – 0.0016 in) |
| | **All models** | |
| Oversizes | +0.5 mm (0.02 in) and +1.0 mm (0.04 in) | |
| Gudgeon pin outside diameter | 11.995 – 12.0 mm (0.4722 – 0.4724 in) | |
| Service limit | 11.96 mm (0.471 in) | |
| Gudgeon pin hole diameter | 12.0 – 12.006 mm (0.4724 – 0.4727 in) | |
| Service limit | 12.07 mm (0.475 in) | |

### Piston rings
                                                                 
| | |
|---|---|
| End gap (fitted) ................................................ | 0.10 – 0.30 mm (0.004 – 0.012 in) |
| Service limit ..................................................... | 0.6 mm (0.024 in) |

### Crankshaft assembly

| | |
|---|---|
| Maximum runout ................................................ | 0.10 mm (0.004 in) |
| Maximum big-end side clearance ......................... | 0.60 mm (0.024 in) |
| Maximum big-end radial clearance ....................... | 0.07 mm (0.003 in) |
| Maximum connecting rod twist ............................ | 0.20 mm (0.008 in) |
| Maximum connecting rod bend ............................ | 0.20 mm (0.008 in) |
| Small-end eye bore diameter .............................. | 16.003 – 16.014 mm (0.6300 – 0.6305 in) |
| Service limit ..................................................... | 16.05 mm (0.632 in) |

### Kickstart

| | |
|---|---|
| Drive pinion bore diameter ................................. | 16.016 – 16.034 mm (0.630 – 0.631 in) |
| Service limit ..................................................... | 16.08 mm (0.633 in) |
| Kickstart shaft diameter .................................... | 15.939 – 15.984 mm (0.627 – 0.629 in) |
| Service limit ..................................................... | 15.92 mm (0.626 in) |
| Idler pinion bore diameter ................................. | 12.016 – 12.034 mm (0.473 – 0.474 in) |
| Service limit ..................................................... | 12.08 mm (0.475 in) |
| Gearbox output shaft diameter ........................... | 11.976 – 11.994 mm (0.471 – 0.472 in) |
| Service limit ..................................................... | 11.93 mm (0.469 in) |
| Driven pinion bore diameter ............................... | 22.020 – 22.041 mm (0.867 – 0.868 in) |
| Service limit ..................................................... | 22.08 mm (0.869 in) |

*Note clutch centre bush diameter*

### Clutch

| | |
|---|---|
| Type .............................................................. | Wet, multiplate |
| Friction plate: | |
|     Thickness .................................................... | 3.12 – 3.28 mm (0.123 – 0.130 in) |
|     Service limit ............................................... | 3.0 mm (0.118 in) |
|     Plate tang to drum slot clearance ................. | 0.30 – 0.65 mm (0.012 – 0.025 in) |
|     Service limit ............................................... | 0.9 mm (0.035 in) |
| Plate maximum warpage ................................... | 0.30 mm (0.012 in) |
| Drum: | |
|     Centre bore diameter .................................... | 22.0 – 22.021 mm (0.8661 – 0.8669 in) |
|     Service limit ............................................... | 22.03 mm (0.8673 in) |
|     Centre bush outside diameter ......................... | 21.965 – 21.980 mm (0.8647 – 0.8653 in) |
|     Service limit ............................................... | 21.94 mm (0.8637 in) |

### Gearbox

| | | |
|---|---|---|
| Type: | | |
|     AE and AR50 ............................................. | 5-speed, constant mesh | |
|     AE and AR80 ............................................. | 6-speed, constant mesh | |
| Gear ratios (no of teeth): | | |
|     1st .......................................................... | 3.308:1 (43/13) | |
|     2nd ......................................................... | 2.111:1 (38/18) | |
|     3rd .......................................................... | 1.545:1 (34/22) | |
|     4th .......................................................... | 1.240:1 (31/25) | |
|     5th .......................................................... | 1.074:1 (29/27) | |
|     6th (80 only) ............................................ | 0.966:1 (28/29) | |
| Primary reduction ratio ..................................... | 3.619:1 (76/21) | |
| Secondary reduction ratio: | | |
|     AE and AR50 ............................................. | 3.769:1 (49/13) | |
|     AE80 ....................................................... | 2.867:1 (43/15) | |
|     AR80 ....................................................... | 3.071:1 (43/14) | |
| Overall drive ratio (in top gear): | | |
|     AE and AR50 ............................................. | 14.649:1 | |
|     AE80 ....................................................... | 10.023:1 | |
|     AR80 ....................................................... | 10.736:1 | |
| Primary drive maximum backlash ........................ | 0.14 mm (0.005 in) | |
| Spring free length: | | |
|     Gearchange shaft arm .................................. | 21.5 mm (0.846 in) | |
|     Service limit ............................................... | 22.6 mm (0.889 in) | |
| | **Early A1 models** | **Later models** |
|     Set levers ................................................. | 25.6 mm (1.008 in) | 22.6 mm (0.890 in) |
|     Service limit ............................................... | 26.9 mm (1.059 in) | 24.0 mm (0.945 in) |
| Selector fork claw end thickness ......................... | 4.9 – 5.0 mm (0.193 – 0.197 in) | |
| Service limit ..................................................... | 4.7 mm (0.185 in) | |
| Pinion groove width .......................................... | 5.05 – 5.15 mm (0.199 – 0.203 in) | |
| Service limit ..................................................... | 5.25 mm (0.206 in) | |
| Selector fork drum locating pin diameter .............. | 4.9 – 5.0 mm (0.193 – 0.197 in) | |
| Service limit ..................................................... | 4.85 mm (0.191 in) | |
| Gearchange drum groove width .......................... | 5.05 – 5.20 mm (0.199 – 0.205 in) | |
| Service limit ..................................................... | 5.25 mm (0.206 in) | |

Pinion to shaft and pinion to bush clearances:
    01 pinion ....................................................................... 0.022 – 0.058 mm (0.0008 – 0.0023 in)
    Service limit ............................................................... 0.16 mm (0.0063 in)
    02, 03, 04, D5 and D6 pinions ..................................... 0.032 – 0.078 mm (0.0012 – 0.0031 in)
    Service limit ............................................................... 0.17 mm (0.0067 in)

## Torque wrench settings

| Component | kgf m | lbf ft |
|---|---|---|
| Cylinder head nuts | 2.2 | 16.0 |
| Spark plug | 2.8 | 20.0 |
| Clutch hub bolt | 2.5 | 18.0 |
| Clutch spring bolts | 0.25 | 22.0 lbf in |
| Flywheel generator rotor nut | 3.0 | 22.0 |
| Neutral switch | 1.2 | 8.7 |
| Gearbox oil drain plug | 2.0 | 14.5 |
| Engine mounting nuts | 2.6 | 19.0 |

**US model specifications are as above except for the following information:**

### Engine
Compression ratio:
    AR50 ......................................................................... 8.0:1
    AR80 ......................................................................... 7.7:1

### Cylinder barrel
Standard bore:
    AR80 ......................................................................... 49.005 – 49.020 mm (1.929 – 1.930 in)

### Cylinder head (also applies to UK AE80 B1 and AR80 C1)
Distortion limit:
    Inner surface ............................................................. 0.05 mm (0.002 in)
    Outer surface ............................................................ 0.10 mm (0.004 in)
    Difference ................................................................. 0.15 mm (0.006 in)

### Piston
Outside diameter:
    AR50 ......................................................................... 38.965 – 38.980 mm (1.5340 – 1.5346 in)
Piston to bore clearance:
    AR50 ......................................................................... 0.030 – 0.040 mm (0.0012 – 0.0016 in)
    AR80 ......................................................................... 0.035 – 0.045 mm (0.0014 – 0.0018 in)

### Gearbox
Type:
    AR50 ......................................................................... 6-speed, constant mesh
Gear ratios (no of teeth):
    1st ............................................................................ 3.308:1 (43/13)
    2nd ........................................................................... 2.111:1 (38/18)
    3rd ............................................................................ 1.545:1 (34/22)
    4th ............................................................................ 1.240:1 (31/25)
    5th ............................................................................ 1.074:1 (29/27)
    6th ............................................................................ 0.966:1 (28/29)
Secondary reduction ratio:
    AR50 ......................................................................... 3.539:1 (46/13)
    AR80 ......................................................................... 2.867:1 (43/15)
Overall drive ratio (in top gear):
    AR50 ......................................................................... 12.372:1
    AR80 ......................................................................... 10.023:1
Set lever spring free length ............................................. 22.6 mm (0.889 in)
Service limit .................................................................. 24.0 mm (0.945 in)

## 1  General description

Kawasaki AE and AR50 and 80 motorcycles are fitted with a basic single-cylinder, air-cooled, 2-stroke engine of familiar design amongst small Japanese motorcycles. This unit employs vertically split crankcases which house both the crankshaft assembly and the gear clusters. The built-up crankshaft has full flywheels and runs on two journal ball bearings which are housed in the crankcase, one each side of the flywheels. Both small-end and big-end bearings are of the caged needle roller type.

The induction system is of the piston port and reed valve design; where the induction of fuel/air mixture into the crankcase is timed by the reciprocating piston skirt and where an even flow of this mixture is maintained by the reed valve. This valve also serves to reduce the possibility of any blow-back of the combustible gases.

Engine lubrication is by means of the Kawasaki Superlube system which takes the form of a crankshaft driven oil pump drawing oil from a separate frame-mounted oil tank and distributing it to the various working parts of the engine. The pump is interconnected to the throttle so that satisfactory lubrication is achieved at all times, thereby corresponding to the requirements of both engine speed and throttle opening. Lubrication for the gearbox and primary transmission is provided by an oil reservoir shared between the two interconnected assemblies.

A flywheel generator is mounted on the left-hand end of the crankshaft. The clutch is mounted on the right-hand end of the gearbox input shaft. Engine starting is by kickstart, drive being passed from the kickstart shaft pinion to the crankshaft via an idler pinion to the clutch drum and then from the clutch drum onto the crankshaft-mounted primary drive pinion.

## 2   Operations with engine/gearbox unit in the frame

1   It is not necessary to remove the engine/gearbox unit from the frame in order to carry out the following service operations:

    a)   Removal and fitting of the cylinder head and barrel
    b)   Removal and fitting of the piston and rings
    c)   Removal and fitting of the carburettor and reed valve
    d)   Removal and fitting of the oil pump
    e)   Removal and fitting of the clutch and primary drive pinion
    f)   Removal and fitting of the gearchange components (external)
    g)   Removal and fitting of the kickstart assembly
    h)   Removal and fitting of the flywheel generator and ignition pick-up
    i)   Removal and fitting of the gearbox sprocket
    j)   Removal and fitting of the neutral indicator switch

## 3   Operations with the engine/gearbox unit removed from the frame

1   Certain operations can be accomplished only if the complete engine unit is removed from the frame. This is because it is necessary to separate the crankcase to gain access to the parts concerned. These operations include:

    a)   Removal and fitting of the crankshaft assembly and main bearings
    b)   Removal and fitting of the gearbox shaft assemblies, gearbox bearings and gearchange components (internal)

## 4   Removing the engine/gearbox unit from the frame

1   Raise the machine to an acceptable working height, using a purpose-built lift or by building a strong platform. Prevent the machine from rolling off its stand by locking on the front brake with a large rubber band around the lever and twistgrip.
2   Position a container of sufficient capacity beneath the engine and remove the gearbox oil drain plug and sealing washer from the underside of the crankcase.
3   Detach each sidepanel and remove the seat. Isolate the battery from the electrical system by disconnecting one of its leads, to prevent shorting of exposed contacts. It is advisable to service the battery at this stage, see Chapter 6. Trace the leads from the flywheel generator to the block connector in front of the battery and separate the two halves of the connector.
4   Observing the necessary fire precautions, turn the fuel tap lever to 'Off' and unclip the feed pipe from the tap stub. On AE models, release the filler cap breather pipe from the handlebars, remove the oil tank filler cap and detach the slotted cover from the front of the tank whilst noting the fitted position of the spacers and washers retained by its two securing screws.
5   Remove the retaining rubber from the rear of the fuel tank. Ease the tank rearwards off its mounting rubbers and place it in safe storage away from any source of naked flame or sparks. Renew any damaged mounting rubbers.
6   Remove the two air inlet duct retaining bolts and pull the duct up to clear the air filter housing. Remove the three housing securing screws, release the inlet hose clamp and ease the housing clear of the frame and carburettor. Renew any split or perished sealing rings. It is advisable to check the filter element, see Routine Maintenance.

7   Unscrew the mixing chamber top from the carburettor and pull the throttle valve assembly clear of the engine. Wrap this assembly in polythene to prevent contamination. Pull the suppressor cap from the spark plug and loosen the plug.
8   Remove the exhaust pipe to cylinder retaining nuts. On AE models, release the expansion chamber to silencer clamp, remove the chamber to frame retaining bolt and ease the chamber forward to clear the machine. On AR models, remove the system to frame retaining bolts and remove the complete system. Note the fitted position of any mounting spacers, renew flattened spring washers or damaged seals and mounting rubbers.
9   Remove the oil pump cover from the right-hand crankcase cover. Release the oil feed pipe from the pump and plug its end with a bolt of the appropriate size. Bend back the small retaining clip from the pump pulley to release the cable nipple. Free the cable inner from the pulley, unscrew the adjuster from the crankcase cover and pull the cable clear of the engine.
10  On AR models, remove the two pump retaining screws and the retaining plate. Pull the pump out of its housing, noting the fitted position of the oil feed dowel and O-rings. Grip the tachometer cable retaining ring and turn the pump to free it from the cable. Upon removal, protect the pump from ingress of dirt.
11  Disconnect the clutch cable from the handlebar lever. Move to the engine, release the cable adjuster locknut, unscrew the adjuster from the crankcase and pull it up and sideways to release the cable from the adjuster location. Pushing the clutch lever upwards will aid this operation. Release the cable from the lever.
12  Note the fitted position of the gearchange lever on its shaft and remove the lever. Remove the gearbox sprocket cover. Place a length of clean rag beneath the final drive chain, remove the chain split link, run the chain off the gearbox sprocket and allow its ends to rest on the rag.
13  Remove the engine mounting bolt retaining nuts and washers. Note the pipe retaining bracket fitted to the top bolt. Protect the front and lower frame tubes with rag or tape. Support the engine, remove the mounting bolts and ease the unit clear of the frame from the left-hand side.

4.6 Detach the air inlet duct

## 5   Dismantling the engine/gearbox unit: preliminaries

1   Clean the unit by using paraffin and a paintbrush or toothbrush. Do not contaminate electrical components, and mask off points where the paraffin can enter the engine internals. Do not use petrol because of the fire risk.
2   Position the dried unit on a clean work surface. Gather some clean rag, a pen and paper and some small containers suitable for storing engine parts.

3   Before commencing work, carefully read the appropriate procedure. Great force is seldom required to remove a component; if in doubt, re-check the procedure.

4   An impact driver is essential for screw removal although it may be possible to use a crosshead screwdriver fitted with a T-handle. Any damaged screws should be renewed or replaced with Allen-headed screws.

### 6   Dismantling the engine/gearbox unit: removing the cylinder head, barrel and piston

1   Remove and examine the spark plug, see Routine Maintenance. Avoid distortion of the cylinder head by slackening its retaining nuts evenly and in a diagonal sequence. Remove the nuts, lightly tap the head around its base with a soft-faced hammer to free it and remove it. Do not lever the head from position.

2   Release the carburettor securing clamp and pull the carburettor from the inlet stub. Disconnect the oil feed pipe from the stub and mask the pipe end.

3   Using a soft-faced hammer, tap the supported areas of finning around the cylinder barrel base to free it from the crankcase. Ease the barrel clear of the crankcase mouth. If the crankcase halves are not to be separated, pack the mouth with clean rag to prevent broken piston rings from falling into the crankcase. Remove the barrel, taking care to support the piston as it leaves the bore.

4   Use small screwdriver to carefully remove one of the gudgeon pin circlips. Press the pin free of the piston, remove the piston and small-end bearing. Discard both circlips. If the pin is tight, warm the piston to free it. A rag soaked in hot water and wrapped around the piston should suffice. Do not tap the pin from position with the connecting rod unsupported, otherwise the rod may bend.

### 7   Dismantling the engine/gearbox unit: removing the oil pump (AE models) and right-hand crankcase cover

1   With the exception of disconnecting the tachometer drive cable, follow the instructions given in Section 4 to remove the oil pump from AR models.

2   Remove the kickstart lever, having noted its fitted position. Making a template of the cover will provide a reference to the fitted position of its retaining screws. Remove these screws evenly and in a diagonal sequence and pull the cover free of its locating dowels.

### 8   Dismantling the engine/gearbox unit: removing the clutch, gearchange components (external), kickstart and primary drive pinion

1   Remove the centre piece from the clutch spring retaining plate. Lock the crankshaft by passing a close-fitting bar through the small-end eye. Place wooden blocks between the crankcase mouth and bar. Remove the clutch retaining bolt.

2   Avoid distortion of the spring retaining plate by slackening its retaining bolts evenly and in a diagonal sequence. Remove the bolts, plate and springs. Remove the clutch hub, friction, plain and end plates as an assembly. Remove the thick thrust washer, the drum and centre bush, kickstart driven pinion and the thin thrust washer.

3   Release the gearchange shaft arm from the gearchange drum and pull the shaft out of the crankcase. Note the fitted position of the gear and neutral set levers, remove their retaining screw and remove the levers and spring.

4   Using a stout pair of pliers, pull the kickstart return spring from its end location in the crankcase and restrain it as it unwinds. Turn the kickstart shaft fully anti-clockwise and pull the complete assembly from the crankcase. Retain the thrust washer which may remain in the crankcase.

5   Remove the kickstart idler pinion retaining circlip. Remove the thrust washer and pinion.

6   Remove the primary drive pinion retaining circlip. Using two large screwdrivers, carefully lever the pinion and damper off the crankshaft. Remove the centre bush and thrust washer. Retain the oil pump drive pin in the crankshaft end.

6.4 Remove the gudgeon pin circlips

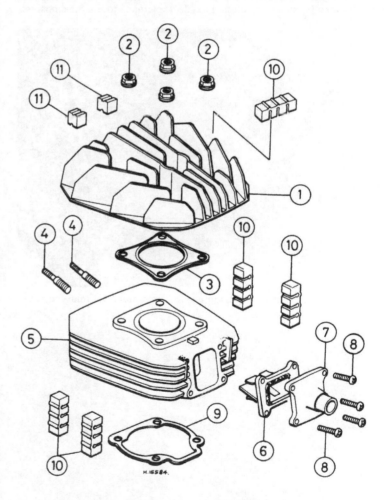

Fig. 1.1 Cylinder head and barrel

| | | | |
|---|---|---|---|
| 1 | Cylinder head | 7 | Inlet stub |
| 2 | Nut – 2 off | 8 | Screw – 4 off |
| 3 | Cylinder head gasket | 9 | Cylinder base gasket |
| 4 | Stud – 2 off | 10 | Damping rubber – 5 off |
| 5 | Cylinder barrel | 11 | Damping rubber – 10 off |
| 6 | Reed valve block | | |

## 9  Dismantling the engine/gearbox unit: removing the flywheel generator and gearbox sprocket

1   Remove the generator cover from the left-hand side of the engine. Lock the crankshaft by passing a close-fitting bar through the small-end eye. Place wooden blocks between the crankcase mouth and bar. Alternatively, fit a strap wrench around the rotor. Remove the rotor retaining nut.
2   Use a flywheel puller of the type shown to remove the rotor. This tool can be purchased quite cheaply from most dealers. If possible, remove the dowel pin from the crankshaft taper.
3   Disconnect the electrical lead from the neutral indicator switch. Release the stator leads from the crankcase and note the fitted position of the stator. This is particularly important on US models, which do not have timing marks; if necessary scratch or lightly punch your own alignment marks. Remove the stator retaining screws and remove the stator.
4   Remove the gearbox sprocket retaining circlip, the sprocket, spacer and O-ring.

## 10  Dismantling the engine/gearbox unit: separating the crankcase halves

1   Support the crankcase on the work surface, screw heads uppermost. Make a template to provide a reference to the fitted position

9.1 Lock the crankshaft and remove the generator rotor retaining nut

9.2 A flywheel puller can be used to remove the rotor

of each screw. Remove the screws evenly whilst working in a diagonal sequence.
2   Using a soft-faced hammer, tap around the crankcase joint to break the seal. Protect the crankshaft end by refitting the generator rotor retaining nut. Hold the crankcase clear of the work surface by its left-hand half and tap sharply down on the crankshaft end with a heavy hammer to drive the shaft out of its location. Check the crankcase halves separate evenly. Do not attempt to lever the halves apart. If unsuccessful, obtain Kawasaki tool no 57001-1098.

## 11  Dismantling the engine/gearbox unit: removing the gearbox components and crankshaft

1   Note the fitted position of each selector fork and remove the forks with their shafts. Remove the input and output shafts and pinions as one assembly. Remove the gearchange drum.
2   Firmly support the right-hand crankcase half on wooden blocks so that the crankshaft end is clear of the work surface. Place a heavy soft-metal drift against the shaft end and strike it with a heavy hammer to drift the shaft down and clear of the crankcase. Do not use excessive force; if unsuccessful, consult your local Kawasaki dealer for advice.

## 12  Examination and renovation: general

1   Before examination, clean each engine part thoroughly in a petrol/paraffin mix whilst observing the necessary fire precautions.
2   Examine each casting for cracks or damage. A crack will require specialist repair. Check each part for wear against the figures given in Specifications. If in doubt, play safe and renew.
3   Repair stripped or badly worn threads by using a thread insert. Refer to your dealer for this service. A screw extractor should be used to remove sheared studs or screws. If unsuccessful or in doubt, consult a professional engineering firm.

## 13  Examination and renewal: crankcase and gearbox oil seals

1   When the crankcase oil seals on a two-stroke engine become worn they admit air into the crankcase and this weakens the incoming fuel/air mixture thereby causing uneven running and difficulty in starting.
2   Carefully examine each seal, especially its lip. Any damage or hardening of a seal will necessitate its immediate renewal. It is advisable to renew the seals as a set, as a matter of course, whilst the engine is dismantled.
3   Remove each seal by prising it from position with the flat of a screwdriver. Avoid damaging the alloy of the seal housing during removal and fitting. A new seal must face in the right direction and enter its housing squarely. Push it in as far as possible by hand and then use a socket and hammer to tap it fully home whilst supporting the casing around the seal housing.

10.0 Carefully remove each oil seal

## 14 Examination and renewal: crankshaft main bearings and gearbox bearings

1   Vibration felt through the footrests and an audible rumble from the bottom end of the engine will signify failure of the crankshaft main bearings. Wash all oil from the bearings and check for play or roughness. Failure will be obvious. When renewing the main bearings order them from an authorized Kawasaki dealer to ensure that the latest strengthened parts are obtained.

2   Remove an unserviceable bearing by detaching its oil seal or retaining plate and applying heat to the crankcase casting. Heat will cause the alloy casting to expand and free the bearing. The safest way to apply heat is to place the casting in an oven set at 80 – 100°C. Warpage may occur if excessive or localised heat is used. Support the casting and use a hammer and socket to tap the bearing free whilst keeping it square to its housing.

3   Before fitting a new bearing, clean its housing and any oil feed drilling. Reheat and support the casting. Use a socket against the bearing outer race to drift the bearing home.

4   Any bearing or bush fitted in a blind housing will require the use of a slide hammer to remove it. Needle bearings will be destroyed upon removal. Use a shouldered drift when fitting a bush or needle bearing. The drift must be a good fit within the bush or bearing to prevent collapse of the same.

5   If the main bearings have remained on the crankshaft, it is advisable to return the shaft to your Kawasaki dealer who will supply and fit new bearings. Any excessive force used on the crankshaft will knock it out of alignment.

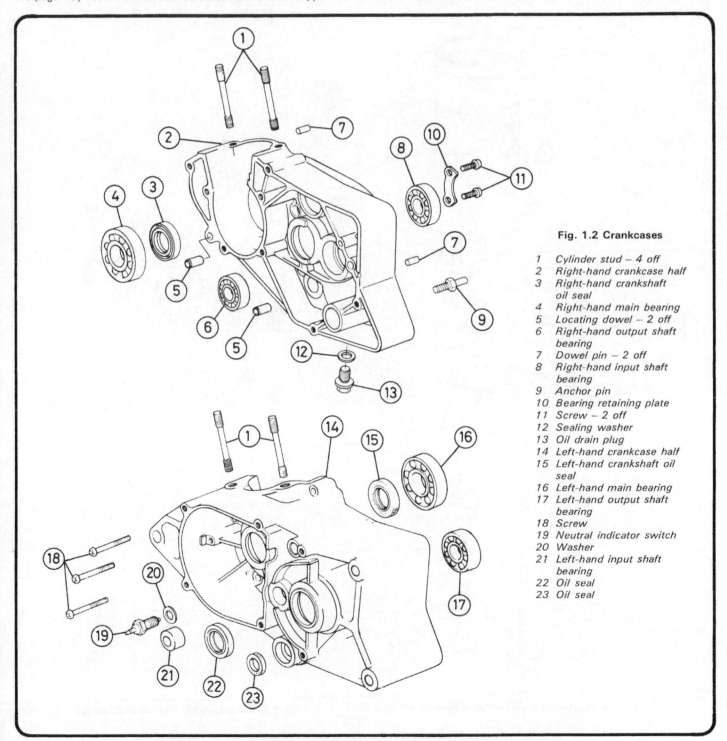

**Fig. 1.2 Crankcases**

1   Cylinder stud – 4 off
2   Right-hand crankcase half
3   Right-hand crankshaft oil seal
4   Right-hand main bearing
5   Locating dowel – 2 off
6   Right-hand output shaft bearing
7   Dowel pin – 2 off
8   Right-hand input shaft bearing
9   Anchor pin
10  Bearing retaining plate
11  Screw – 2 off
12  Sealing washer
13  Oil drain plug
14  Left-hand crankcase half
15  Left-hand crankshaft oil seal
16  Left-hand main bearing
17  Left-hand output shaft bearing
18  Screw
19  Neutral indicator switch
20  Washer
21  Left-hand input shaft bearing
22  Oil seal
23  Oil seal

## 15 Examination and renovation: crankshaft assembly

1  Big-end failure is characterised by a pronounced knock which will be most noticeable when the engine is worked hard. The usual causes of failure are normal wear, or a failure of the lubrication supply. In the case of the latter, big-end wear will become apparent very suddenly, and will rapidly worsen.

2  Check for wear with the crankshaft set in the TDC (top dead centre) position, by pushing and pulling the connecting rod. No discernible movement will be evident in an unworn bearing, but care must be taken not to confuse end float, which is normal, and bearing wear.

3  Obtain a dial gauge and a set of feeler gauges. Set the crankshaft on V-blocks positioned on a completely flat surface or between centres on a lathe. Refer to the accompanying figures and to the service limits in Specifications. Check the crankshaft for runout and big-end side and radial clearances. Check the connecting rod for distortion.

4  Push the small-end bearing into the connecting rod eye and fit the gudgeon pin through the bearing. Hold the rod steady and feel for movement between it and the pin. If movement is felt, renew the pin, bearing or connecting rod as necessary, so no movement exists. The small-end eye bore diameter must not exceed 16.05 mm (0.632 in) or the pin diameter be less than 11.96 mm (0.471 in). Renew the bearing if its roller cage is cracked or worn.

5  Do not attempt to dismantle the crankshaft assembly, this is a specialist task. If the service limits are exceeded or a fault found, return the assembly to a Kawasaki dealer who will supply a new or service-exchange item.

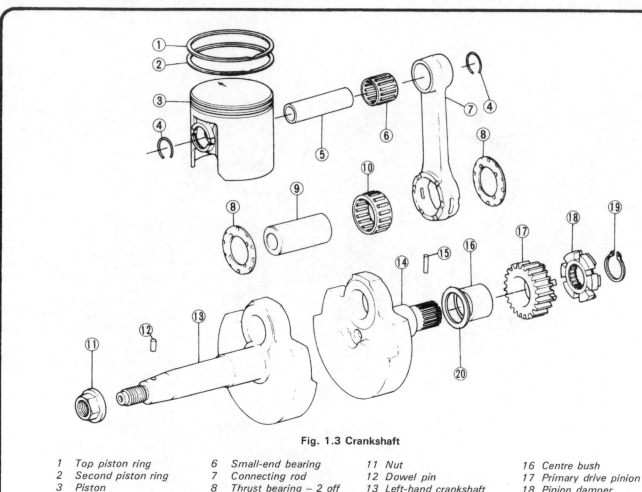

**Fig. 1.3 Crankshaft**

| | | | |
|---|---|---|---|
| 1  Top piston ring | 6  Small-end bearing | 11  Nut | 16  Centre bush |
| 2  Second piston ring | 7  Connecting rod | 12  Dowel pin | 17  Primary drive pinion |
| 3  Piston | 8  Thrust bearing – 2 off | 13  Left-hand crankshaft | 18  Pinion damper |
| 4  Circlip – 2 off | 9  Crank pin | 14  Right-hand crankshaft | 19  Circlip |
| 5  Gudgeon pin | 10  Big-end bearing | 15  Oil pump drive pin | 20  Thrust washer |

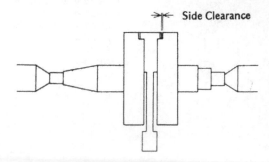

Fig. 1.4 Measuring big-end side clearance

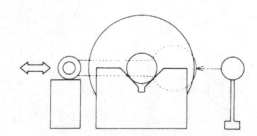

Fig. 1.5 Measuring big-end radial clearance

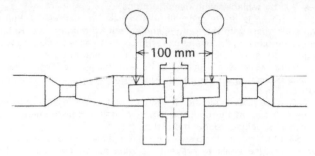

Fig. 1.6 Measuring connecting rod twist

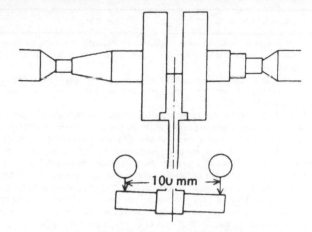

Fig. 1.7 Measuring connecting rod bend

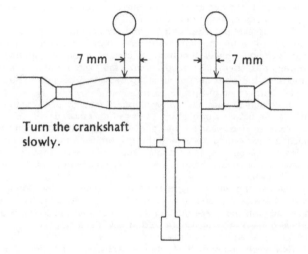

Turn the crankshaft slowly.

Fig. 1.8 Measuring crankshaft runout

## 16 Examination and renovation: decarbonising

1    Refer to Routine Maintenance for details of decarbonising the engine and exhaust system.

## 17 Examination and renovation: cylinder head

1    Check that the cylinder head fins are not clogged with oil or road dirt, otherwise the engine will overheat. If necessary, use a degreasing agent and brush to clean between the fins. Check that no cracks are evident, especially in the vicinity of the spark plug or stud holes.
2    Check the condition of the thread in the spark plug hole. If it is damaged an effective repair can be made using a Helicoil thread insert. This service is available from most Kawasaki agents. Always use the correct plug and do not overtighten, see Routine Maintenance.
3    Leakage between the head and barrel will indicate distortion. Check the head by placing a straight-edge across several places on its mating surface and attempting to insert a 0.10 mm (0.004 in) feeler gauge between the two. This applies to all except US and later UK 80 models, for which see accompanying figure.
4    Remove excessive distortion by rubbing the head mating surface in a slow, circular motion against emery paper placed on plate glass. Start with 200 grade paper and finish with 400 grade paper and oil. Do not remove an excessive amount of metal. If in doubt consult a Kawasaki agent.
5    Note that most cases of cylinder head distortion can be traced to unequal tensioning of the cylinder head securing nuts or by tightening them in the incorrect sequence.

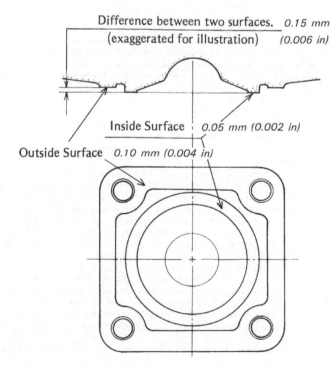

Difference between two surfaces. *0.15 mm*
(exaggerated for illustration)    *(0.006 in)*

Inside Surface    *0.05 mm (0.002 in)*

Outside Surface    *0.10 mm (0.004 in)*

Fig. 1.9 Checking the cylinder head for distortion – US AR80, UK AE80 B1 and AR80 C1 models

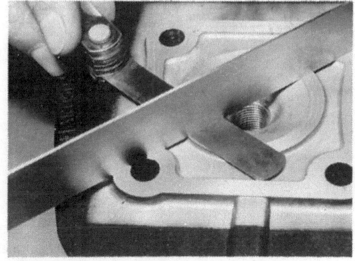

17.3 Check cylinder head distortion

## 18 Examination and renovation: cylinder barrel

1   The usual indication of a badly worn cylinder barrel and piston is piston slap, a metallic rattle that occurs when there is little or no load on the engine.
2   Clean all dirt from between the cooling fins. Carefully remove the ring of carbon from the bore mouth so that bore wear can be accurately assessed.
3   The US AR80 model is fitted with a barrel which cannot be honed or rebored as the bore is plated with an electrically-deposited coating to provide a hard-wearing but lightweight bearing surface. This means that if either the barrel or the piston are found to be worn to the specified service limit or beyond, they must be renewed.
4   Examine the bore for scoring or other damage, particularly if broken rings are found. Damage will necessitate reboring and a new piston. A satisfactory seal cannot be obtained if the bore is not perfectly finished.
5   There will probably be a lip at the uppermost end of the cylinder bore which marks the limit of travel of the top of the piston ring. The depth of the lip will give some indication of the amount of bore wear that has taken place even though the amount of wear is not evenly distributed.
6   The most accurate method of measuring bore wear is by the use of a cylinder bore DTI (Dial Test Indicator) or a bore micrometer. Measure at the points shown in the accompanying figure, in-line with the gudgeon pin axis and at 90° to it, six measurements in all. Refer to Specifications for the bore wear limit.
7   If any of the measurements obtained exceed the service limit for that size of bore (if oversized, the bore wear limit is found by adding 0.1 mm (0.004 in) to the diameter that the cylinder was bored out to; if this is unknown, measure the bore at the bottom of the barrel, where it will be relatively unworn) the barrel must be rebored and an oversized piston and rings must be fitted. If the rebore will enlarge the bore diameter to more than 1 mm over the maximum standard value the cylinder must be renewed. Also if there is more than 0.05 mm (0.002 in) difference between any two measurements, the barrel must be rebored.
8   After a rebore, chamfer the edges of each port to prevent the piston rings catching on them and breaking. This requires careful use of a scraper with fine emery paper to finish. Take care not to damage the bore.
9   Measure piston to bore clearance either by direct measurement of the piston and bore diameters and subtraction or by measurement of the gap with a feeler gauge. Piston diameter must be measured 5 mm (0.20 in) from the base of its skirt and at right-angles to the gudgeon pin hole. If the clearance measured exceeds that given in Specifications, a new piston or rebore and new piston is required.
10  Refer to the preceding Section and check the barrel to head mating surface for distortion.

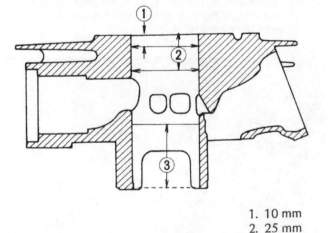

1. 10 mm
2. 25 mm
3. 40 mm

Fig. 1.10 Cylinder bore wear measurement points

## 19 Examination and renovation: piston and piston rings

1   Disregard the existing piston and rings if a rebore is necessary,
they will be replaced with oversize items.
2   Each ring is brittle and will break easily if overstressed. Pull the ring ends apart with the thumbs and ease it from its groove. Refer to the accompanying figure and use strips of tin to ease free a ring which is gummed in its groove. Note the fitted position of each ring.
3   Piston wear usually occurs at its skirt, taking the form of vertical streaks or scoring on the thrust side. There may be some variation in skirt thickness.
4   Measure piston diameter 5 mm (0.20 in) from the base of its skirt, at right-angles to the gudgeon pin hole. If the measurement is less than that specified, renew the piston.
5   Slight scoring of the piston can be removed by careful use of a fine swiss file. Use chalk to prevent clogging of the file teeth and the subsequent risk of scoring. Bad scoring will indicate the need for piston renewal.
6   Any build-up of carbon in the ring grooves can be removed by using a section of broken ring, the end of which has been ground to a chisel edge. If the ring grooves are unevenly worn or the ring locating pegs are loose or worn, renew the piston.
7   The gudgeon pin should be a firm press fit in the piston. Check for scoring on the bearing surfaces of each part and where damage or wear is found, renew the part affected. The pin circlip retaining grooves must be undamaged; renew the piston rather than risk damage to the bore through a circlip becoming detached. Note the service limits given in Specifications.
8   Discoloured areas on the working surface of each piston ring indicate the blow-by of gas and the need for renewal.
9   Measure ring wear by inserting each ring into part of the bore which is unworn and measuring the gap between the ring ends with a feeler gauge. If the measurement exceeds that given in Specifications, renew the ring. Use the piston crown to locate the ring squarely in the bore.
10  Remove the wear ridge at the top of the bore before fitting new rings, otherwise the top ring will strike the ridge and break. This should be included in the 'glaze busting' process carried out to break down surface glaze on the used bore so that new rings can bed in. Refer to the Kawasaki agent for this service.
11  Refit each ring correctly in its previously noted position, checking that it presses easily into its groove and that its ends locate correctly over the peg. As with part worn rings, the end gap of new rings must be measured. If necessary, carefully use a needle file to enlarge the gap.

## 20 Examination and renovation: gearbox components

1   Examine each gear pinion for chipped or broken teeth and dogs. Renew each damaged item. When dismantling and reassembling each shaft assembly, refer to the accompanying figure. Gear selection problems will occur if the pinions, thrust washers or circlips are incorrectly fitted. To reduce confusion, reassemble as soon as possible and make rough sketches, where necessary. Renew worn or distorted washers and circlips. Note the pinion groove width limit given in Specifications.
2   With the gearbox shafts dismantled, inspect the shaft and pinion splines for wear and hairline cracks, renewing each component as necessary. If a pinion has a bushed centre and the bush is overworn, do not immediately reject the pinion but ask a competent engineer if the bush can be renewed. Note the pinion to shaft clearances given in Specifications and the associated kickstart and clutch information.
3   Clean the gearbox sprocket thoroughly and examine it for hooked or broken teeth and wear of the centre spline. Do not renew the sprocket on its own but renew both sprockets and the chain to prevent rapid wear resulting from the running together of old and new parts.
4   Check for scoring on the bearing surface of the selector fork ends, bores or drum locating pins. Check for cracks around the bore edges and at the base of the fork arms. Where the fork ends and locating pins have worn beyond the limits given in Specifications, renew the fork in question.
5   Position the selector fork shaft on a sheet of plate glass and check for straightness by attempting to insert a feeler gauge beneath it. A bent shaft will cause gearchange problems. Check the shaft and gearchange drum bearing surfaces for scoring and wear and renew where necessary. Check the drum groove width does not exceed the limit given in Specifications.

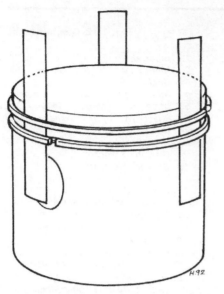

Fig. 1.11 Method of freeing gummed piston rings

19.4 Measure piston diameter

19.9 Measure piston ring wear (end gap)

19.11 Locate the piston ring ends each side of the peg

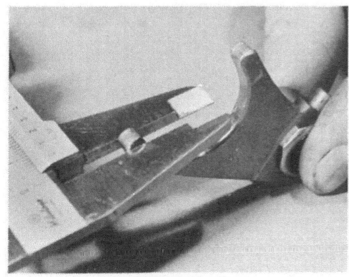

20.4a Measure selector fork claw end thickness ...

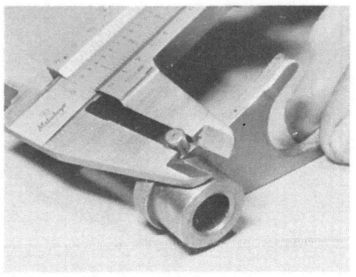

20.4b ... and measure the drum locating pin diameter

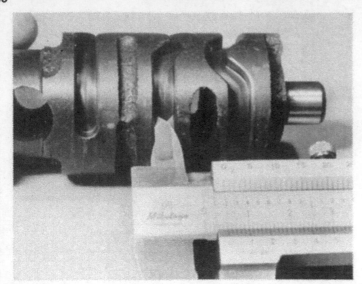

20.5 Measure the gearchange drum groove width

## Fig. 1.12 Gearbox shaft

1 Input shaft
2 Input shaft 5th gear pinion
3 Circlip – 6 off
4 Input shaft 3rd/4th gear pinion
5 Input shaft 6th gear pinion – 6 speed models
6 Spacer (8.5 mm) – 5-speed models
7 Input shaft 2nd gear pinion
8 Circlip
9 Gearbox sprocket
10 Spacer (10 mm)
11 O-ring
12 Output shaft
13 Output shaft 2nd gear pinion
14 Output shaft 6th gear pinion – 6-speed models,
   sliding dog – 5-speed models
15 Thrust washer – 2 off
16 Output shaft 4th gear pinion
17 Output shaft 3rd gear pinion
18 Output shaft 5th gear pinion
19 Output shaft 1st gear pinion

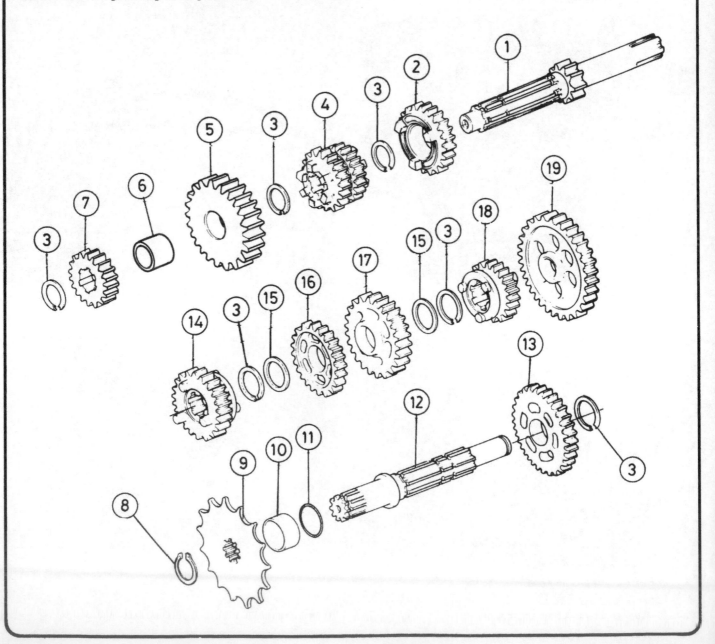

## 21 Examination and renovation: gearchange mechanism

1   Check for wear between the gearchange shaft arm pawls, set lever and pins of the drum. Check each pin is a firm fit in its location. If necessary, use an impact driver to remove the drum end plate retaining screw and release the pins. Renew all the pins as a set. Check the pin retaining holes for ovality and if necessary, renew the drum.

2   Where excessive wear is found between any two bearing surfaces of the mechanism, renew the component(s) concerned. Renew any spring which has become fatigued or taken a set beyond the service limits given in Specifications.

3   Position the gearchange shaft on a sheet of plate glass and check it for straightness by attempting to insert a feeler gauge beneath it. A bent shaft will cause gearchange problems.

4   Later A1 models were fitted with a modified set of gearchange mechanism components, as shown in this manual. On early models if any of these components require renewal, take the machine's engine and frame numbers with you when ordering the replacement parts (to be certain of obtaining the correct items, take the old parts as patterns). Note that items 1, 12, 13, and 15 from Fig. 1.13 are only interchangeable as a set.

## 22 Examination and renovation: clutch and primary drive

1   Clean the clutch components in a petrol/paraffin mix whilst observing the necessary fire precautions.

2   Overworn friction plates will cause clutch slip. Renew each plate if its thickness is less than the specified limit. Check the plate tangs and drum slots for indentations caused by clutch chatter. If slight, damage can be removed with a fine file. If damage is excessive or the specified tang to slot clearance exceeded, renewal is necessary.

3   Check all plates for distortion by laying each one on a sheet of plate glass and attempting to insert a feeler gauge beneath it. Refer to Specifications for maximum allowable warpage.

4   Examine each plain plate for scoring and signs of overheating in the form of blueing. Remove slight damage to the plate tangs and hub slots with a fine file, otherwise renewal is necessary. Relubricate all plates before assembly.

5   Examine the spring retaining plate and centre piece, the hub and the end plate for cracks, overheating and excessive distortion. Look for hairline cracks around the base of each end plate spring location and around the centre boss of the hub.

6   Any wear or scoring of the thrust washers and drum centre bush will necessitate their renewal. Renew the drum or bush if worn beyond the specified limits. Pay special attention to the slots in the drum centre which accommodate the drive dogs of the kickstart driven pinion, these have been known to fracture. If these components require renewal refer to Chapter 7, Section 3 for details of the modified driven pinion and clutch outer drum.

7   The operating mechanism can be pulled from the right-hand crankcase cover. Check each part for wear and renew as necessary. Renew the O-ring if flattened or damaged. Regrease the mechanism during assembly. Check the ball bearing is secure in the spring retaining plate centre piece. Note that the clutch cable fitted to later A1 models had a plastic sleeve enclosing the lower end nipple; this modification necessitated the fitting of a modified release lever.

8   Examine the teeth of both the primary drive and driven pinions for excessive wear or damage. Both sets of teeth will have worn in unison and should be renewed as a matched pair, the driven pinion being part of the clutch drum.

9   Renew the drive pinion centre bush and thrust washer if scored or excessively worn. Examine the pinion damper for damage or deterioration and renew if necessary; it must be a firm fit over the crankshaft splines and pinion dogs. Renew the retaining circlip if distorted.

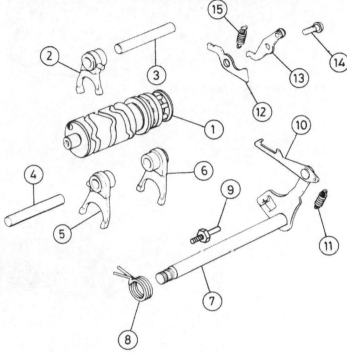

**Fig. 1.13 Gearchange mechanism**

| 1 | Gearchange drum | 9 | Spring anchor |
|---|---|---|---|
| 2 | Selector fork | 10 | Gearchange shaft arm |
| 3 | Selector fork shaft | 11 | Spring |
| 4 | Selector fork shaft | 12 | Gear set lever |
| 5 | Selector fork | 13 | Neutral set lever |
| 6 | Selector fork | 14 | Screw |
| 7 | Gearchange shaft | 15 | Spring |
| 8 | Return spring | | |

22.2a Measure friction plate thickness ...

22.2b ... and check plate tang to slot clearance

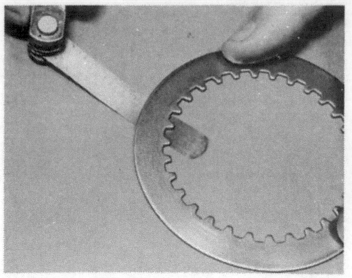

22.3 Check each clutch plate for distortion

22.6 Examine the clutch centre bush and kickstart driven pinion

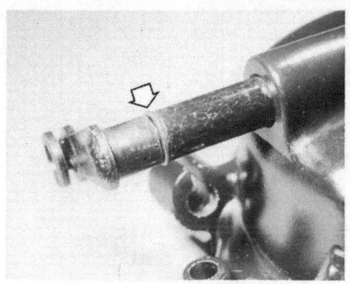

22.7 Examine the O-ring of the clutch operating mechanism

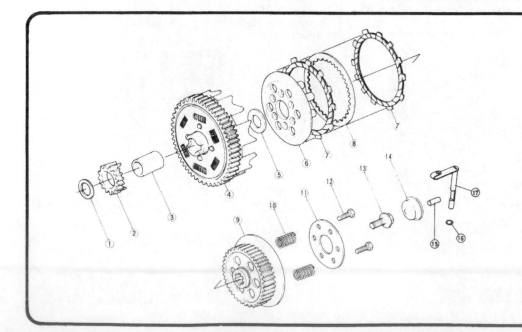

### Fig. 1.14 Clutch

1 Thick thrust washer
2 Kickstart driven pinion
3 Clutch centre bush
4 Clutch drum
5 Thick thrust washer
6 Clutch hub
7 Friction plate
8 Plain plate
9 End plate
10 Spring – 6 off
11 Retaining plate
12 Bolt – 6 off
13 Centre bolt
14 Centre piece
15 Pushrod
16 O-ring
17 Release lever

## 23 Examination and renovation: kickstart assembly

1 Dismantle the kickstart shaft assembly, cleaning each part and placing it on a clean worksurface in the order of removal. Renew fatigued or broken springs and circlips and obviously worn or damaged parts. If any part of the kickstart lever requires renewal, ensure that the latest strengthened components are obtained from an authorized Kawasaki dealer.

2 The drive pinion teeth will wear in unison with those of the idler and driven pinions. In the case of excessive wear, renew the pinions as a set. Note the shaft and pinion wear limits given in Specifications. Examine the dogs of the driven pinion for hairline cracks.

3 The ratchet gear and drive pinion will wear in unison, necessitating their renewal as a pair. Renew the ratchet stop and guide plates fitted to the crankcase if excessively worn. Use an impact driver to remove the plate retaining screws and coat their threads with locking compound before fitting.

4 Lubricate each shaft component before reassembly and refer to the accompanying figure where necessary. Align the punch marks of the ratchet gear and shaft as shown.

23.4a Fit the thrust washer and circlip each side of the kickstart drive pinion ...

23.4b ... align the punch marks of the shaft and ratchet gear ...

23.4c ... fit the ratchet spring, circlip and thrust washer ...

23.4d ... locate the return spring in the shaft ...

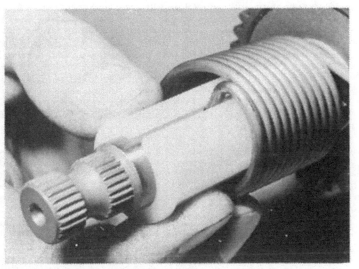

23.4e ... and fit the spring guide

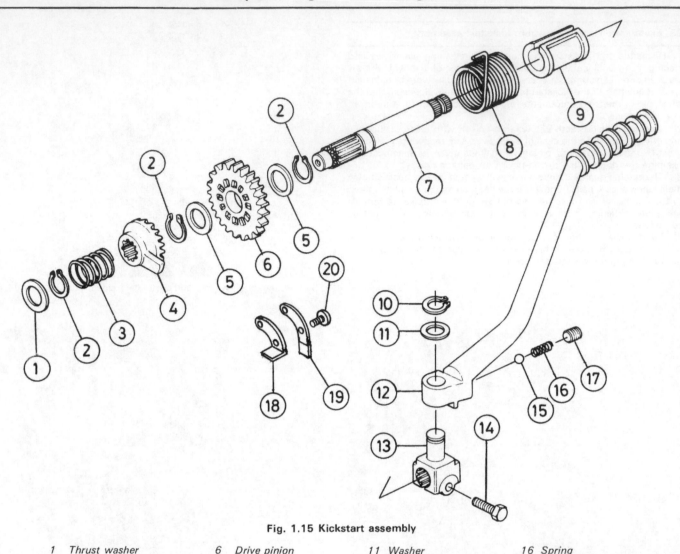

**Fig. 1.15 Kickstart assembly**

| | | | |
|---|---|---|---|
| 1 | Thrust washer | 6 | Drive pinion |
| 2 | Circlip – 3 off | 7 | Kickstart shaft |
| 3 | Ratchet spring | 8 | Return spring |
| 4 | Ratchet gear | 9 | Spring guide |
| 5 | Thrust washer – 2 off | 10 | Circlip |

| | | | |
|---|---|---|---|
| 11 | Washer | 16 | Spring |
| 12 | Kickstart lever | 17 | Grub screw |
| 13 | Lever knuckle | 18 | Ratchet stop plate |
| 14 | Bolt | 19 | Ratchet guide plate |
| 15 | Steel ball | 20 | Screw – 2 off |

## 24 Examination and renovation: oil pump and tachometer drive gear (AR models)

1   The oil pump is effectively a sealed unit, no replacement parts being available. Examination is limited to cleaning the unit body and checking for hairline cracks around its screw holes and their stress points. Check the cable pulley is returned by its spring and renew O-rings as a matter of course.

2   The pump drive pin fits in the crankshaft end. Damage to the pin or the pump drive spigot will necessitate immediate renewal of the component(s) concerned.

3   The tachometer drive fitted to AR models is secured to the pump by two screws. As with the pump, examine the unit body and drive attachments. If the components are separated, renew the drive to pump sealing ring. Before fitting the drive, align the notch in its shaft with the corresponding spigot in the pump.

## 25 Engine reassembly: general

1   Thoroughly clean each component. Avoid damaging mating surfaces when removing old gaskets or sealing compound. Carefully separate gaskets from components with a scalpel or finely honed chisel. Remove stubborn remnants of gasket and sealing compound by

soaking with methylated spirits or a similar solvent. Use a soft brass-wire brush to scrub off non-hardening compound.

2   Place the components on a clean surface near the working area together with all new gaskets and seals and any new parts. Check that nothing is missing.

3   Lay out all necessary tools. Fill an oil can with clean SAE 10W/30 oil; each moving component should be lubricated during assembly. Wipe clean the assembly area.

4   Observe all specified torque and clearance settings during assembly. Renew all damaged fasteners. Where screw heads have been damaged, it is worth considering a set of Allen screws as a more robust replacement.

## 26 Reassembling the engine/gearbox unit: assembling the gear clusters

1   Refer to the accompanying figures and photographs when assembling the gear clusters. Thoroughly clean and lubricate each component before fitting. Seat each circlip properly in its groove aligning the gap between its ends with the base of one of the shaft spline channels, as shown.

2   On UK 50 models (5-speed), note that the 8.5 mm long input shaft spacer (item 6, Fig. 1.12) must not be interchanged with the 10 mm long spacer (item 10) fitted behind the gearbox sprocket.

26.1a Fit the 2nd gear pinion and circlip to the gearbox output shaft ...

26.1b ... followed by the top gear pinion (6-speed) or sliding dog (5-speed), circlip and thrust washer ...

26.1c ... the 4th gear pinion ...

26.1d ... the 3rd gear pinion, thrust washer and circlip ...

26.1e ... the 5th gear pinion ...

26.1f ... and the 1st gear pinion

26.1g Fit the 5th gear pinion and circlip to the input shaft ...

26.1h ... followed by the 3rd/4th gear pinion and circlip ...

26.1i ... the 8.5 mm long spacer (5-speed) or top gear pinion (6-speed) and 2nd gear pinion ...

26.1j ... and the 2nd gear pinion circlip (AE50 shown)

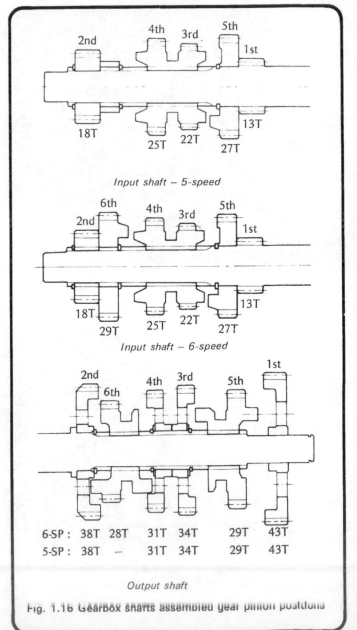

*Input shaft – 5-speed*

*Input shaft – 6-speed*

| | | | | | | |
|---|---|---|---|---|---|---|
| 6-SP : | 38T | 28T | 31T | 34T | 29T | 43T |
| 5-SP : | 38T | – | 31T | 34T | 29T | 43T |

*Output shaft*

Fig. 1.16 Gearbox shafts assembled gear pinion positions

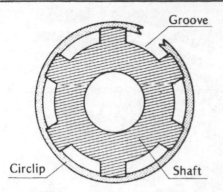

Fig. 1.17 Correct position of a circlip on a splined shaft

27.3a Fit the gearbox components ...

## 27 Reassembling the engine/gearbox unit: refitting the crankshaft and gearbox components

1   If necessary, refit any bearing retaining plate, coating the threads of its screws with locking compound. Lubricate the main and gearbox bearings and grease the lip of each oil seal.
2   Support the right-hand crankcase half on wooden blocks. Position the crankshaft squarely in its location and push sharply down to seat it. If necessary, tap the shaft into position, using a length of thick-walled tube placed over its end. Do not use excessive force and support the flywheels at a point opposite the crankpin, to prevent distortion. If difficulty is experienced, gently heat the crankcase half.
3   Lubricate and fit the gearchange drum, gear clusters, selector forks and shafts. Take care to retain any thrust washers and refer to the accompanying photograph for the correct fitted position of the selector forks.

## 28 Reassembling the engine/gearbox unit: joining the crankcase halves

1   Degrease the mating surface of each crankcase half. Press the two locating dowels into the right-hand half and coat the mating surface of the same half with a thin application of sealing compound.
2   Lower the left-hand crankcase half over the shaft ends, aligning each one with its location. Using hand pressure, press the crankcase halves together, and follow up with a soft-faced hammer, tapping around the crankshaft location to bring the mating surfaces together. Do not use excessive force.
3   Fit the crankcase securing screws in their previously noted locations. Prevent distortion by tightening the screws evenly and in a diagonal sequence. Wipe away excess sealing compound and check the crank and gearbox shafts rotate freely. Investigate any tightness.

## 29 Reassembling the engine/gearbox unit: refitting the gearbox sprocket and flywheel generator

1   Grease the end splines of the gearbox output shaft and the lip of its oil seal. Grease and fit a new O-ring, pushing it through the seal with the sprocket spacer (10 mm long). Fit the gearbox sprocket and its retaining circlip. Check the circlip is a good firm fit.
2   Relocate the stator in its previously noted position and tighten its retaining screws. If a new stator is being fitted, align the stator timing mark with the crankcase pointer shown. Clip the electrical leads to the crankcase and reconnect the neutral indicator switch.
3   Degrease the rotor and crankshaft mating surfaces. Check the dowel pin is firmly fitted in the crankshaft and push the rotor over it. Gently tap the rotor centre with a soft-faced hammer to seat it, lock the crankshaft and tighten the rotor retaining nut to the specified torque loading.

27.3b ... and position the selector forks

28.2 Join the crankcase halves

29.1a Push the O-ring through the oil seal with the gearbox sprocket spacer (10 mm) ...

29.1b ... and fit the sprocket and circlip

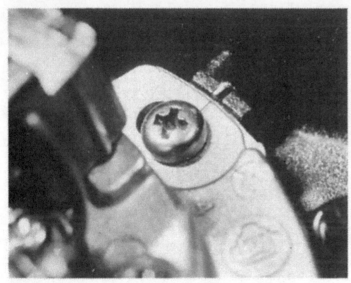

29.2a Align the stator timing mark (UK models only) with the crankcase pointer

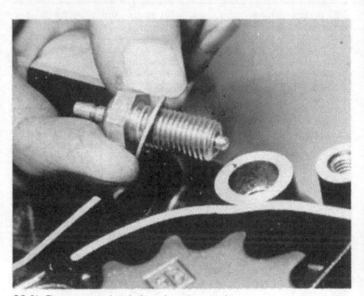

29.2b Fit the neutral switch and sealing washer ...

29.2c ... and reconnect the switch electrical lead

29.3a Check the dowel pin is firmly fitted in the crankshaft taper ...

29.3b ... before pushing the rotor into position

30.1a Fit the thrust washer, centre bush and primary drive pinion

## 30 Reassembling the engine/gearbox unit: refitting the primary drive pinion, kickstart, gearchange components (external) and clutch

1    Check the oil pump drive pin is fitted properly in the crankshaft end. Fit the thrust washer, centre bush and primary drive pinion. Push the pinion damper over the crankshaft splines, aligning it with the dogs of the pinion. Use a suitable socket and hammer to tap the damper fully home. Fit the pinion retaining circlip, checking it is a good firm fit.

2    Fit the thrust washer from the kickstart shaft into its crankcase location. Fit the shaft, turning it clockwise to lock the ratchet stop behind its stop and guide plates. Use a stout pair of pliers to tension the return spring and push its end into the crankcase location.

3    Fit the kickstart idler pinion, thrust washer and circlip. Fit the thin thrust washer, clutch centre bush and kickstart driven pinion.

4    If necessary, refit the gearchange drum pins and end plate, coating the threads of the plate retaining screw with locking compound. Fit the gear and neutral set levers and spring. Apply locking compound to the threads of the lever pivot bolt. Tighten the bolt, check the levers align correctly with the drum and pivot freely.

5    Check tighten the gearchange shaft return spring locating pin in the crankcase. If loose, recoat its threads with locking compound. Grease the splined end of the gearchange shaft and carefully insert it through the crankcase. The legs of the return spring must locate one each side of the locating pin and shaft spigot. Relocate the shaft arm over the gearchange drum.

6    Fit the clutch drum and thick thrust washer. Assemble the hub, friction, plain and end plates and insert the assembly into the drum. Where the friction plates have diagonal rather than straight grooves, note that they must be fitted so that the grooves run in the same direction as the clutch (see Fig. 1.14). Fit the springs, retaining plate and bolts. Avoid distortion of the plate by tightening the bolts evenly and in a diagonal sequence to the recommended torque loading. Lock the crankshaft and tighten the clutch retaining bolt to the recommended torque loading. Fit the centre piece and grease its ball bearing.

30.1b Push the pinion damper fully home and fit the circlip

## 31 Reassembling the engine/gearbox unit: refitting the right-hand crankcase cover and oil pump (AE models)

1    Check all components within the cover and crankcase are properly fitted and lubricated. Grease the splined end of the kickstart shaft and the lip of the cover oil seal.

2    Degrease the cover and crankcase mating surfaces. Press the locating dowels into the crankcase and fit the gasket over them. Fit the cover, tapping around it lightly with a soft-faced hammer to seat it. Fit each screw into its previously noted position. Tighten the screws evenly whilst working in a diagonal sequence. Fit the kickstart lever in its previously noted position and tighten its clamp bolt.

3    On AE models, press the oil feed dowel and its new O-ring into the crankcase cover. Fit a new O-ring over the oil pump boss, align the pump drive and push the pump into its housing. Secure the pump with the two screws and plate.

30.2a Fit the kickstart shaft into the crankcase ...

30.2b ... tension the return spring and locate its end

30.3 Fit the kickstart idler and driven pinions and the clutch centre bush

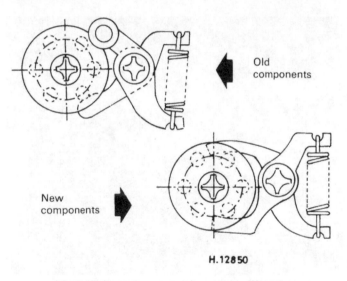

Old components

New components

H.12850

Fig.1.18 Gearchange set levers modification

30.5 Position the gearchange shaft return spring and arm

30.6a Fit the clutch drum and thick thrust washer ...

30.6b ... followed by the hub, friction, plain and end plates ...

30.6c ... fit the springs, retaining plate and bolts ...

30.6d ... and fit and grease the centre piece

31.2 Fit the locating dowels and new gasket before fitting the right-hand cover

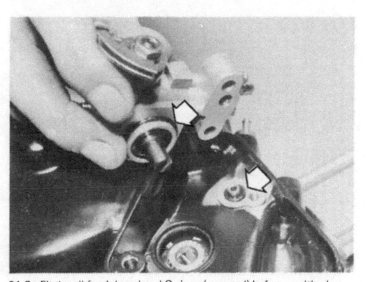

31.3a Fit the oil feed dowel and O-rings (arrowed) before positioning the pump

31.3b Secure the oil pump with the two screws and plate

## 32 Reassembling the engine/gearbox unit: refitting the piston, cylinder barrel and head

1   Lubricate the big-end bearing and pack the crankcase mouth with clean rag. Lubricate and fit the small-end bearing. With the arrow cast in the piston crown facing forward, place the piston over the rod and fit the gudgeon pin. If necessary, warm the piston to aid fitting.

2   Retain the gudgeon pin with new circlips. Check each clip is correctly located; if allowed to work loose it will cause serious damage.

3   Check the piston rings are still correctly fitted. Clean the barrel and crankcase mating surfaces and fit the new base gasket. Lubricate the piston rings and cylinder bore. Lower the barrel over its retaining studs and ease the piston into the bore, carefully squeezing the ring ends together. Do not use excessive force.

4   Remove the rag from the crankcase mouth and push the barrel down onto the crankcase. Clean the barrel to head mating surfaces and fit a new head gasket so that its 'K' mark faces upwards as shown. Fit the head and its retaining nuts. Tighten the nuts evenly and in a diagonal sequence to the specified torque loading.

5   Clean the carburettor to inlet stub mating surfaces. Check the carburettor seals are serviceable and correctly fitted and refit the carburettor. Connecting of the oil feed pipe to the inlet stub should be left until the pump is bled.

32.1a Fit the small-end bearing ...

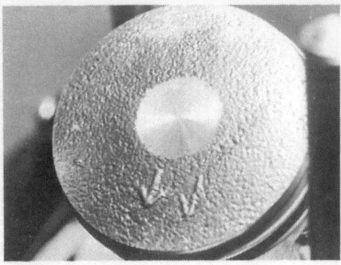

32.1b ... fit the piston with the arrow facing forward . . .

32.1c ... and fit and secure the gudgeon pin with new circlips

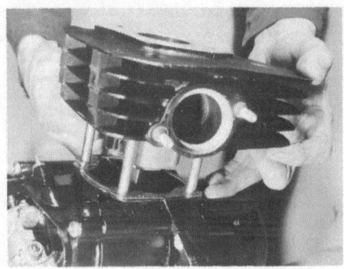

32.3 Carefully lower the cylinder barrel onto its new base gasket

32.4 Fit the cylinder head gasket with the 'K' mark upwards as shown

## 33 Fitting the engine/gearbox unit into the frame

1   Check that no part has been omitted during reassembly. Prepare the machine and ease the engine into the frame from the left-hand side. Align the engine and fit the mounting bolts and nuts. Refit the pipe retaining bracket to the top bolt. Tighten the nuts to the specified torque loading.

2   Reconnect the final drive chain, moving the rear wheel forward if necessary. The spring clip of the link must have the side plate fitted beneath it, be seated correctly and have its closed end facing the direction of chain travel. Refer to Routine Maintenance and check chain adjustment. Refit the gearbox sprocket cover. Refit the gearchange lever in its previously noted position.

3   Reconnect the clutch cable to the crankcase lever and adjuster location, and to the handlebar lever. Refer to Routine Maintenance and check cable adjustment.

4   On AR models, reconnect the oil pump to the tachometer cable. Refer to Section 31 and fit the pump. On all models, reconnect the pump cable to the crankcase cover and pulley, securing the cable nipple with the retaining clip. Unplug and connect the feed pipe to the pump.

5   Refit the exhaust system. Locate a new seal between pipe and barrel and fit all mountings loosely. Tighten the exhaust port nuts first, then the remaining mountings.

6   Check tighten the spark plug; it must have a crush washer fitted. Reconnect the suppressor cap. Route the HT lead so that it cannot chafe on engine or frame.

7   Check the seal inside the carburettor mixing chamber top is intact. Insert the throttle valve assembly into the carburettor and tighten the chamber top. Do not use excessive force. Refer to Routine Maintenance and check throttle cable adjustment.

8   Refit the air filter housing to the frame and carburettor. Check all sealing rings are correctly positioned and the housing to carburettor clamp is fully tightened. Refit the air inlet duct.

9   Refit the fuel tank, checking there is no metal-to-metal contact which will split the tank. Reconnect the fuel feed pipe, turn on the tap and check for leaks. Leaks must be cured otherwise fire may result, causing serious personal injury. Where applicable, check the filler cap and oil tank breather pipes are correctly routed.

10  Reconnect the generator leads. Reconnect the battery. Check all connections are clean and smear the battery terminals with petroleum jelly to prevent corrosion. Connect the battery earth lead to the negative (-) terminal and route its vent pipe clear of the lower frame.

11  Secure the seat and refit both sidepanels. Fit the gearbox oil drain plug and its new sealing washer. Tighten the plug to the specified torque loading. Remove the filler plug and replenish the gearbox with the specified type and quantity of oil, see Routine maintenance. Wait a few minutes before lifting the machine off its stand so that it is vertical to the ground and checking that the oil level coincides with the centre hole of the filler plug dipstick with the plug screwed fully in. Remove or add oil as necessary.

12  Recheck each disturbed component for security. Check each control for smooth movement over its full operating range. Refer to Chapter 2 and bleed the oil pump. Using a syringe, prime the feed pipe to the inlet stub with 2-stroke oil. Reconnect the pipe.

13  Before attempting to start the engine, refer to Routine Maintenance and read the instructions on oil pump adjustment and ignition timing. Obtain the equipment necessary to carry out these procedures.

## 34 Starting and running the rebuilt engine

1   Attempt to start the engine. Some coaxing may be necessary at first but if unsuccessful, check the plug for fouling. If problems persist, consult Fault Diagnosis.

2   Upon starting, run the engine slowly, opening the choke as soon as possible. Any oil used during reassembly will soon burn away. Check oil pump adjustment and ignition timing as soon as the engine is running evenly.

3   Warm the engine thoroughly and check for blowing gaskets and oil leaks. Refit the pump and generator covers. Check all gears select properly and all controls function correctly.

## 35 Taking the rebuilt machine on the road

1   Give the machine time to settle down by treating it gently for the first few miles. If rebored, the engine will have to be run in. This means greater use of the gearbox and a restraining hand on the throttle for at least 500 miles. Keep a light loading on the engine and gradually work up performance until this limit is reached. Where only a new crankshaft is fitted, carry out these recommendations to a lesser extent.

2   If a lubrication failure is suspected, immediately stop and investigate otherwise irreparable engine damage is inevitable.

3   Do not add oil to the petrol. This will only create excess smoke and accelerate the rate of carbon build-up in the combustion chamber and exhaust. The oil pump will provide full lubrication.

4   Do not tamper with the exhaust system or remove the silencer baffle. Doing this will decrease engine performance, as will removing the air filter.

5   After the initial run, allow the engine to cool and check all components for security. Re-adjust any controls which may have settled down.

33.1 Fit the pipe retaining bracket to the top mounting bolt

33.4a On AR models, connect the tachometer cable before fitting the oil pump

33.4b Secure the oil pump cable with the pulley clip

# Chapter 2 Fuel system and lubrication

*Refer to Chapter 7 for information relating to the 1984 on models*

## Contents

## Specifications

**Note:** The following specifications are for UK models only. Owners of US models check with the information at the end of this Section.

### Fuel tank capacity
AE50 and 80 ............................................................ 6.5 lit (1.4 Imp gal) overall, 0.6 lit (0.13 Imp gal) reserve
AR50 and 80 ............................................................ 9.6 lit (2.1 Imp gal) overall, 0.8 lit (0.18 Imp gal) reserve

### Fuel grade ............................................................ Unleaded, or leaded (minimum octane rating 91 RON/RM)

### Carburettor

| | AE50 | AE80 | AR50 | AR80 |
|---|---|---|---|---|
| Make ............ | Mikuni | Mikuni | Mikuni | Mikuni |
| Type ............ | VM14SC | VM18SC | VM14SC | VM18SC |
| Main jet (standard) ............ | 112.5 | 125 (A1) 112.5 (B1) | 110 | 122.5 |
| Needle jet ............ | D9 | 07 | D9 | 06 (A1) 05 (C1) |
| Jet needle ............ | 317-3 | 4M4-3 (A1) 4M8-3 (B1) | 317-3 | 4M4-3 (A1) 4M8-4 (C1) |
| Pilot jet ............ | 17.5 or 15 | 17.5 | 17.5 or 15 | 17.5 |
| Starter jet ............ | 35 | 30 | N/Av | N/Av |
| Pilot screw setting (turns out) ............ | 1 | 1$\frac{3}{4}$ | 1 | 1$\frac{3}{4}$ |
| Throttle valve cutaway ............ | 2.0 | 2.5 | 2.0 | 2.0 (A1) 2.5 (C1) |
| Float valve seat ............ | 1.8 | 1.8 | 1.8 | 1.8 |
| Float height ............ | 23.8 ± 1.0 mm (0.94 ± 0.04 in) | 22.2 ± 1.0 mm (0.87 ± 0.04 in) | 23.8 ± 1.0 mm (0.94 ± 0.04 in) | 22.2 ± 1.0 mm (0.87 ± 0.04 in) |
| Fuel level ............ | 3.5 ± 1.0 mm (0.14 ± 0.04 in) | 1.5 ± 1.0 mm (0.06 ± 0.04 in) | 3.5 ± 1.0 mm (0.14 ± 0.04 in) | 1.5 ± 1.0 mm (0.06 ± 0.04 in) |
| Engine idling speed ............ | 1250 rpm | 1250 rpm | 1250 rpm | 1250 rpm |

**Reed valve**

    Reed bend limit ....................................................

**Air cleaner**

    Element type ....................................................

**Engine lubrication**

    System type ....................................................
    Oil type ....................................................
    Oil capacity:
        AE50 and 80 ....................................................
        AR50 and 80 ....................................................
    Oil pump output (after 3 mins at 2000 rpm, pump pulley fully open):
        AE and AR50 ....................................................
        AE and AR80 ....................................................

**Gearbox lubrication**

    Oil type ....................................................
    Oil capacity ....................................................

**All models**

0.2 mm (0.008 in)

Oiled polyurethane foam

Kawasaki Superlube – Pump fed total-loss oil injection
Good quality air-cooled 2-stroke engine oil

1.3 lit (2.3 Imp pint)
1.2 lit (2.1 Imp pint)

0.8 – 1.0 cc (0.03 Imp fl oz)
2.3 – 2.8 cc (0.09 Imp fl oz)

SAE 10W/30 or SAE 10W/40 SE or SF
0.6 lit (1.06 Imp pint)

**US model specifications are as above except for the following information:-**

**Fuel tank capacity**

    AR50 and 80 ....................................................

9.6 lit (2.5 US gal) overall, 0.8 lit (0.21 US gal) reserve

**Carburettor**

| | AR50 | AR80 |
|---|---|---|
| Make | Mikuni | Mikuni |
| Type | VM16SC | VM18SS |
| Main jet (standard) | 115 | 62.5 |
| | | 60 (high altitude) |
| Needle jet | D8 | O5 |
| Jet needle | 317-3 | 4D44 |
| Pilot jet | 17.5 or 15 | 25 |
| | | 20 (high altitude) |
| Starter jet | N/Av | N/Av |
| Pilot screw setting (turns out) | $1^{1}/_{4}$ | N/App |
| Throttle valve cutaway | 2.0 or 2.5 | 1.5 |
| Float valve seat | 1.8 | N/Av |
| Float height | 23.8 ± 1.0 mm | 19.0 ± 1.0 mm |
| | (0.94 ± 0.04 in) | (0.75 ± 0.04 in) |
| Fuel level | 3.5 ± 1.0 mm | 5.0 ± 1.0 mm |
| | (0.14 ± 0.04 in) | (0.19 ± 0.04 in) |
| Engine idling speed | 1250 rpm | 1300 rpm |

**Engine lubrication**

    Oil capacity ....................................................
    Oil pump output (after 3 mins at 2000 rpm, pump pulley
    fully open):
        AR50 ....................................................
        AR80 ....................................................

1.2 lit (1.27 US qt)


1.4 – 1.7 cc (0.05 US fl oz)
2.3 – 2.8 cc (0.09 US fl oz)

**Gearbox lubrication**

    Oil capacity ....................................................

0.6 lit (0.63 US qt)

## 1  General description

Fuel is gravity fed from the tank to the carburettor float chamber via a three-position tap. Air drawn into the carburettor is filtered through an oil-impregnated foam element.

The point of induction on a two-stroke engine is normally controlled by the piston skirt, which covers and uncovers ports machined in the cylinder bore. On Kawasaki AE and AR models, a supplementary timing system, in the form of a reed valve, is incorporated to enable more efficient induction timing. The reed valve maintains an even flow of the combustion mixture and reduces the possibility of blow-back of the combustible gases, thereby contributing greatly towards an economical and powerful engine.

Engine lubrication is by Kawasaki Superlube. Oil is gravity fed from the tank to a pump which is crankshaft driven and interconnected by cable to the throttle twistgrip, thus the amount of oil pumped varies according to engine speed and throttle setting.

Pumped oil is fed directly to the inlet tract where it mixes with the incoming mixture charge being drawn to the crankcase. The oil is deposited to lubricate the bearings and the cylinder bore and piston. Residual oil enters the combustion chamber with the incoming mixture and is burnt.

Lubrication for the transmission components is by oil contained within the gearbox casing, isolated from the working parts of the engine proper.

## 2  Fuel tank: removal and refitting

1   Observing the necessary fire precautions, turn the tap lever to 'Off' and unclip the feed pipe from the tap stub. On AE models, release the filler cap breather pipe from the handlebars, remove the oil tank filler cap and detach the slotted cover from the front of the tank whilst noting the fitted position of the spacers and washers retained by its two securing screws.

2    On all models, release the retaining rubber from the rear of the tank, having removed the seat. Ease the tank rearwards off its mounting rubbers and place it in safe storage, away from any naked flame or sparks. Renew any damaged mounting rubbers.

3    Reverse the removal procedure to fit the tank. If necessary, wipe the mounting rubbers with petrol to ease their installation. Secure the tank and check for metal-to-metal contact which might split the tank.

4    Reconnect the fuel feed pipe, turn on the tap and check for leaks. Leaks must be cured otherwise fire may result, causing serious personal injury. Where applicable, check the filler cap and oil tank breather pipes are correctly routed.

---

### 3    Fuel tap: removal, refitting and curing of leaks

1    Fuel can leak from either the tap to tank joint or tap lever joint. If check-tightening the component fails to effect a cure, then proceed as follows.

2    Leakage at the tap to tank joint will be caused by a defective O-ring. Observe the necessary fire precautions and drain the tank by detaching the fuel feed pipe from the carburettor, placing its end in a clean, sealable metal container and turning on the tap. Detach the tap, taking care not to damage its filter stacks. Remove the defective O-ring.

3    If necessary, clean each stack by rinsing in clean petrol. Remove stubborn contamination by gentle brushing with a used toothbrush or similar item soaked in petrol. Protect the eyes from spray-back from the brush.

4    A holed stack should be renewed as it will allow sediment into the tap body. Clean the tap and tank mating surfaces and fit a new O-ring. Refit the tap, reconnect the pipe, refill the tank and check for leaks.

5    Leakage of the lever joint will be caused by a defective O-ring. Drain the tank and remove the lever by releasing its plate retaining screws. Clean the lever and its location in the tap body. Renew the O-ring. Refit the lever, check its operation, refill the tank and check for leaks.

---

### 4    Fuel feed pipe: examination

1    The thin-walled synthetic rubber pipe is a push-on type and need only be replaced if it has gone hard or split. Renew its retaining clips, if fatigued.

---

### 5    Carburettor: removal and refitting

1    Refer to Section 2 and remove the fuel tank and seat. Detach the right-hand sidepanel. Remove the two air inlet duct retaining bolts and pull the duct up to clear the air filter housing. Remove the three housing securing screws, release the inlet hose clamp and ease the housing clear of the frame and carburettor. Renew any split or

perished seals.

2    Unscrew the mixing chamber top from the carburettor and pull the throttle valve assembly clear of the chamber. Thread the carburettor drain pipe clear of its guide. Release the securing clamp and pull the carburettor from the inlet stub.

3    During fitting, check all seals are correctly located. Air drawn in through a joint will badly affect the engine performance. Clean the carburettor to inlet stub mounting surfaces and refit the carburettor. Reroute the drain pipe. Check the seal inside the carburettor mixing chamber top is intact. Carefully insert the throttle valve assembly into the carburettor and tighten the chamber top. Do not use excessive force. Refer to Routine Maintenance and check throttle cable adjustment.

4    Refit the air filter housing, air inlet duct, right-hand sidepanel, fuel tank and seat. Remember to check for fuel leaks before riding the machine. Leaks must be cured otherwise fire may result, causing serious personal injury.

---

### 6    Carburettor: dismantling, examination, renovation and reassembly

1    Cover an area of work surface with clean paper. This will prevent components placed upon it from becoming contaminated or lost.

2    Remove the float chamber retaining screws. If necessary, tap around the chamber to body joint with a soft-faced hammer to free the chamber. Remove the float pivot pin and detach the twin floats. Displace the float needle, it is very small and easily lost.

2.3 Fit each fuel tank mounting rubber correctly

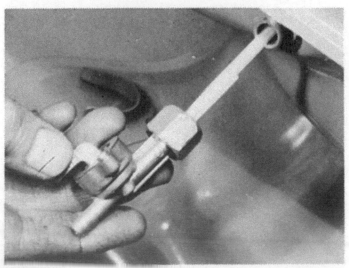

3.2 Take care not to damage the fuel tap filter stacks during removal

6.3a Check the carburettor mouth seals are correctly fitted ...

5.3b ... before pushing the carburettor over the inlet stub

5.3c Carefully insert the throttle valve into the mixing chamber

5.4 Check the inlet hose is securely clamped

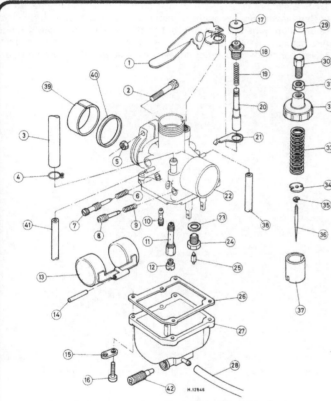

**Fig. 2.1 Carburettor UK 80 models (50 similar)**

| | | | |
|---|---|---|---|
| 1 | Choke operating lever | 25 | Float needle |
| 2 | Bolt | 26 | Float chamber gasket |
| 3 | Fuel pipe | 27 | Float chamber |
| 4 | Pipe clip | 28 | Overflow pipe |
| 5 | Nut | 29 | Rubber cover |
| 6 | Spring | 30 | Adjusting screw |
| 7 | Throttle stop screw | 31 | Locknut |
| 8 | Pilot screw | 32 | Mixing chamber top |
| 9 | Spring | 33 | Return spring |
| 10 | Pilot jet | 34 | Needle retaining plate |
| 11 | Needle jet | 35 | Circlip |
| 12 | Main jet | 36 | Jet needle |
| 13 | Float | 37 | Throttle valve |
| 14 | Float pivot pin | 38 | Vent pipe* |
| 15 | Washer – 2 off* | 39 | Insulator |
| 16 | Screw – 4 off | 40 | Seal |
| 17 | Dust seal | 41 | Vent pipe* |
| 18 | Choke plunger housing | 42 | Drain plug* |
| 19 | Spring | | |
| 20 | Choke plunger | | * where fitted |
| 21 | Lever plate | | |
| 22 | Carburettor body | | Note: pilot screw located at |
| 23 | Washer | | front of carburettor body on |
| 24 | Float needle seat | | 50 models |

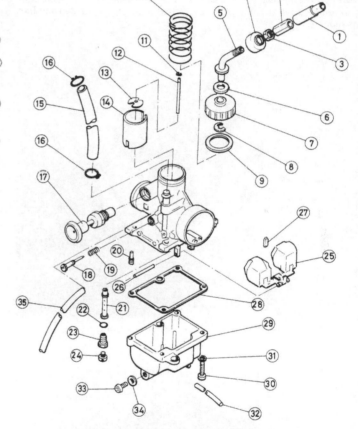

**Fig. 2.2 Carburettor – US AR80 model**

| | | | |
|---|---|---|---|
| 1 | Rubber cover | 19 | Spring |
| 2 | Adjuster | 20 | Pilot jet |
| 3 | Locknut | 21 | Needle jet |
| 4 | Dust cover | 22 | O-ring |
| 5 | Cable sleeve | 23 | Jet holder |
| 6 | Washer | 24 | Main jet |
| 7 | Mixing chamber top | 25 | Float |
| 8 | Circlip | 26 | Pivot pin |
| 9 | Sealing ring | 27 | Float needle |
| 10 | Return spring | 28 | Gasket |
| 11 | Circlip | 29 | Float chamber |
| 12 | Jet needle | 30 | Screw – 4 off |
| 13 | Needle retaining plate | 31 | Spring washer – 4 off |
| 14 | Throttle valve | 32 | Overflow pipe |
| 15 | Fuel pipe | 33 | Drain plug |
| 16 | Clip – 2 off | 34 | Sealing washer |
| 17 | Choke plunger assembly | 35 | Vent pipe |
| 18 | Throttle stop screw | | |

3   Unscrew the float needle seat and washer, the main jet, needle jet and pilot jet. When removing any jet, use a close fitting screwdriver of the correct type otherwise damage will occur.

4   Note the pilot screw (where fitted) and throttle stop screw settings by counting the number of turns required to screw them fully in until seating lightly; this will make it easier to 'retune' the carburettor after reassembly. Remove both screws with their springs.

5   Knock back the tab washer and remove the choke assembly. It is not necessary to remove the drain plug from the float chamber except for seal renewal.

6   Before examination, thoroughly clean each part in clean petrol, using a soft nylon brush to remove stubborn contamination and a compressed air jet to blow dry. Avoid using rag because lint will obstruct jet orifices. Do not use wire to clear blocked jets, this will enlarge the jet and increase petrol consumption; if an air jet fails use a soft nylon bristle. Observe the necessary fire precautions and wear eye protection against blow-back from the air jet.

7   Renew any distorted or cracked casting. Renew all O-rings and gaskets. Replace fatigued or broken springs and flattened spring washers.

8   Wear of the float needle takes the form of a groove around its seating area; renew if worn. Check for similar wear of the needle seat and renew if worn. Check the needle end pin is spring loaded and free to move.

9   Check the floats for damage and leakage. Renew if damaged, it is not advisable to attempt a repair.

10  Examine the choke assembly, renewing any worn or damaged parts. Renew hardened drain or fuel feed pipes.

11  Wear of the throttle valve will be denoted by polished areas on its external diameter, causing air leaks which weaken the mixture and produce erratic slow running. Examine the carburettor body for similar wear and renew each component as necessary.

12  Examine the jet needle for scratches or wear along its length and for straightness. If necessary, dismantle the valve assembly by pushing the return spring against the mixing chamber top and releasing the throttle cable from the valve. Remove the needle retaining plate or clip and remove the needle from the valve.

13  Renew the seal within the mixing chamber top if damaged. The throttle return spring must be free of fatigue or corrosion.

14  Before assembly, clean all parts and place them on clean paper in a logical order. Do not use excessive force during reassembly, it is easy to shear a jet or damage a casting.

15  Reassembly is a reverse of dismantling. If in doubt, refer to the accompanying figures or photographs. Seat the pilot and throttle stop screws lightly before screwing out to their previously noted settings. Alternatively set the pilot screw as specified and refer to Routine Maintenance for carburettor adjustment to set the throttle stop screw.

6.12a Remove the retaining clip from within the throttle valve ...

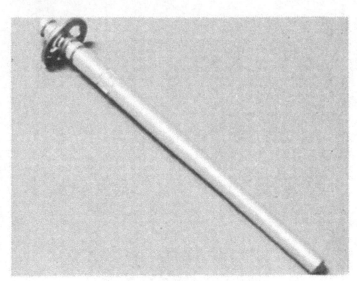

6.12b ... withdraw and examine the jet needle (AE50 shown)

6.13 Examine the seal within the mixing chamber top

6.15a Lock the choke assembly in position with the tab washer (AE50 shown)

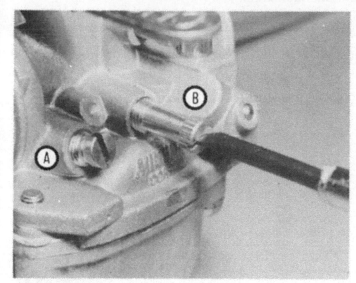

6.15b Reset the pilot screw (A) and throttle stop screw (B) ...

6.15c ... fit the pilot jet ...

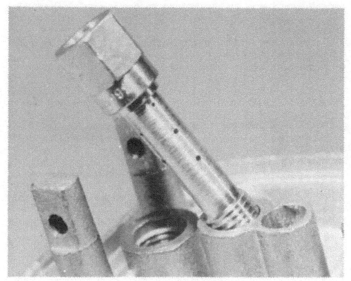

6.15d ... the needle jet ...

6.15e ... the main jet ...

6.15f ... the float needle seat and sealing washer ...

6.15g ... the float needle ...

6.15h ... and retain the floats with the pivot pin (AE50 shown)

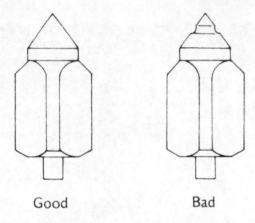

Good                    Bad

Fig. 2.3 Example of float needle wear

## 7   Carburettor adjustment and exhaust emissions: general note

In some countries legal provision is made for describing and controlling the types and levels of toxic emissions from motor vehicles.

In the USA exhaust emission legislation is administered by the Environmental Protection Agency (EPA) which has introduced stringent regulations relating to motor vehicles. The Federal law entitled Clean Air Act, specifically prohibits the removal (other than temporary) or modification of any component incorporated by the vehicle manufacturer to comply with the requirements of the law. The law extends the prohibition to any tampering which includes the addition of components, use of unsuitable replacement parts or maladjustment of components which allows the exhaust emissions to exceed the prescribed levels. Violations of the provisions of this law may result in penalties of up to $10 000 for each violation. It is strongly recommended that appropriate requirements are determined and understood prior to making any change to or adjustments of components in the fuel, ignition, crankcase breather or exhaust systems.

To help ensure compliance with the emission standards, some manufacturers have fitted to the relevant systems fixed or pre-set adjustment screws as anti-tamper devices. In most cases this is restricted to plastic or metal limiter caps fitted to the carburettor pilot adjustment screw, which allows normal adjustment only within narrow limits. Occasionally the pilot screw may be recessed and sealed behind a small metal blanking plug, or locked in position with a thread-locking compound, which prevents no mal adjustment.

It should be understood that none of the various methods of discouraging tampering actually prevents adjustment, nor, in itself, is no adjustment an infringement of the current regulations. Maladjust-

ment, however, which results in the emission levels exceeding those laid down, is a violation. It follows that no adjustments should be made unless the owner feels confident that he can make those adjustments in such a way that the resulting emissions comply with the limits. For all practical purposes a gas analyser will be required to monitor the exhaust gases during adjustment, together with EPA data of the permissible hydrocarbon and CO levels. Obviously, the home mechanic is unlikely to have access to this type of equipment or the expertise required for its use, and, therefore, it will be necessary to place the machine in the hands of a competent motorcycle dealer who has the equipment and skill to check the exhaust gas content.

For those owners who feel competent to carry out correctly the various adjustments, specific information relating to the anti-tamper components fitted to the machines covered in this manual is given in the relevant Sections of this Chapter.

## 8   Carburettor: settings

1   The predetermined jet sizes, throttle valve cutaway and needle position should not require modification. Check with Specifications for standard settings. Carburettor wear occurs very slowly; therefore, if a sudden fault in engine performance occurs, check all other main systems before suspecting the carburettor.
2   The standard fitted position of the jet needle clip is indicated by the suffix number of the needle number identification. For example, 317-3 indicates the clip should be fitted in the 3rd groove down from the needle top.
3   Incorrect float height, leaking seals and defective air filter and exhaust system will all result in bad engine performance.
4   If possible, ride the machine for approximately 5 miles and stop the engine without letting it tick over. Remove the spark plug and check its electrodes and insulator for condition and colour as defined in Routine Maintenance. This will give a good indication of any fault in carburation.
5   As a guide, up to $\frac{1}{8}$th throttle is controlled by the pilot jet, $\frac{1}{8}$ to $\frac{1}{4}$ by the throttle valve cutaway, $\frac{1}{4}$ to $\frac{3}{4}$ by the needle position and from $\frac{3}{4}$ to full by the main jet size. These are approximate divisions which are not clear cut, there being an amount of overlap between the stages.
6   The pilot screw setting is specified. If engine dies at low speed, suspect a blocked pilot jet. Refer to Routine Maintenance for carburettor adjustment.
7   Settings will be affected by fitting non-standard air filters, exhaust systems, etc. Refer to the equipment manufacturer for advice.

## 9   Carburettor: checking fuel level and float height

1   Flooding of the carburettor or excessive mixture weakness will indicate incorrect float height. Remove the carburettor, invert it and remove the float chamber with gasket. Renew this gasket if damaged.
2   Place the inverted carburettor on a flat and level surface. Measure the distance from the body gasket surface to the furthest surface of the float as shown in the accompanying photograph and compare the reading obtained with the figure given in Specifications. If necessary, alter the setting by bending the small tongue sited between the floats.
3   Checking the fuel level is more accurate but on early models (without float chamber drain plugs), involves removing the carburettor to substitute a modified float chamber. The modification is to cut off the brass overflow stand pipe inside the chamber so that a length of plastic tube can be attached to the vent on the float chamber underside. The special tool (part number 57001-1017) is then attached to the other end of the tube and the carburettor is refitted to the machine. On later models with drain plugs another special tool is required which has an adaptor suitable for the drain plug thread. If these tools cannot be obtained, the machine should be taken to an authorised Kawasaki dealer for the fuel level to be checked. If the tools are available, proceed as follows.
4   With the carburettor positioned dead vertical and the gauge held alongside it (see accompanying figure) fuel introduced into the float chamber should fill the gauge to a level below the chamber to carburettor body mating surface. The distance between the mating surface and the fuel in the gauge (the fuel level) should correspond to that specified. If not, refer to the preceding paragraphs and check the float height.

9.2 Check the carburettor float height (AE50 shown)

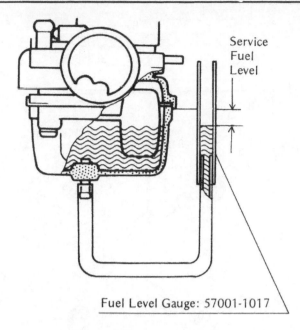

Fuel Level Gauge: 57001-1017

Fig. 2.4 Measuring the fuel level

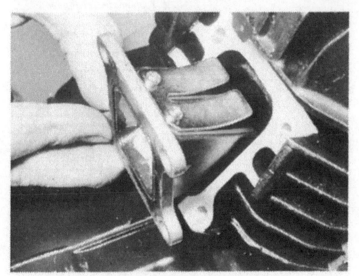

11.1 Remove the inlet stub to expose the reed valve

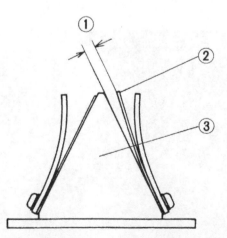

Fig. 2.5 Measuring the amount of reed valve reed bend

| 1 | Measurement | 3 | Holder |
|---|---|---|---|
| 2 | Reed | | |

## 10 Carburettor: adjustment

1    Refer to Routine Maintenance for details of this operation.

## 11 Reed valve: removal, examination and refitting

1    Remove the carburettor. Remove the inlet stub retaining screws. Remove the stub and reed valve. Renew the valve if its mating surface seals or the rubber coating of its body have become damaged.
2    After a considerable mileage, the valve reeds may lose some springiness, causing their performance to suffer. Check for obvious damage, such as cracked reeds. Refer to the accompanying figure and measure the amount of reed bend. If the measurement exceeds 0.2 mm (0.008 in) renew the valve.
3    Before fitting the valve and inlet stub retaining screws, coat their threads with locking compound. Tighten the screws evenly and in a diagonal sequence to avoid distorting the valve body.

## 12 Air filter element: removal, examination, cleaning and refitting

1    Refer to Routine Maintenance for details of this operation.

## 13 Engine lubrication system: general maintenance

1    An adequate level of oil must be maintained in the tank; the warning light will indicate a low level. Use only the oil specified.
2    Regularly check oil feed and delivery pipes for splitting or perishing. All connections must be free from leaks, which will eventually cause loss of lubrication and subsequent engine seizure.
3    Refer to Routine Maintenance for oil pump adjustment if incorrect lubrication is suspected.

## 14 Oil pump: removal and refitting

1    The pump is a sealed unit; in the event of failure it must be renewed. Remove the pump cover from the right-hand crankcase cover. Release the oil feed pipe from the pump and plug its end with a bolt of the appropriate size. Bend back the small retaining clip from the pump pulley to release the cable nipple. Free the cable inner from the pulley.

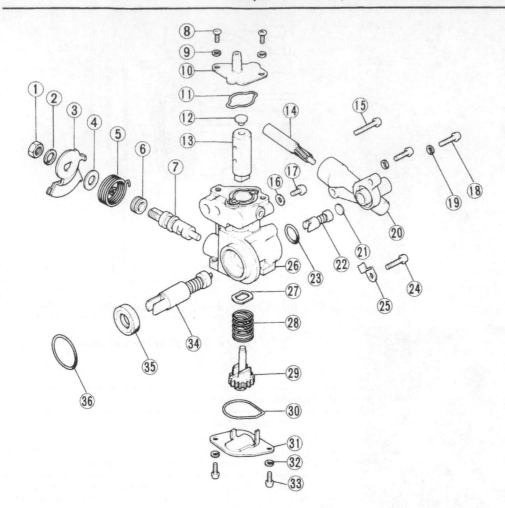

**Fig. 2.6 Oil pump**

1   Nut
2   Spring washer
3   Pump pulley
4   Washer
5   Return spring
6   Ring
7   Operating shaft
8   Screw – 2 off
9   Spring washer – 2 off
10  Top cover
11  O-ring
12  Spacer
13  Plunger
14  Tachometer driven gear
15  Screw
16  Sealing washer
17  Bleed screw
18  Screw
19  Spring washer
20  Tachometer drive housing
21  Washer
22  Tachometer drive gear
23  O-ring
24  Screw
25  Retaining plate
26  Oil pump
27  Spring seat
28  Spring
29  Plunger
30  O-ring
31  End cap
32  Spring washer – 2 off
33  Screw – 2 off
34  Driveshaft
35  Oil seal
36  O-ring

2   Remove the two pump retaining screws and the retaining plate. Pull the pump from its housing, noting the fitted position of the oil feed dowel and O-rings. On AR models, grip the tachometer cable retaining ring and turn the pump to free it from the cable.

3   Before fitting the pump, check its drive is undamaged and clean its mating surface with the crankcase. Renew the pump boss and oil feed dowel O-rings if flattened or damaged. On AR models, reconnect the pump to the tachometer cable.

4   Align the pump drive and push the pump into its housing. Fit the retaining screws and plate. Reconnect the cable to the pulley, securing its nipple with the retaining clip. Unplug and connect the feed pipe to the pump. Refer to Routine Maintenance and carry out cable adjustment. Refer to the following Section and bleed the pump of air. Check for leaks and refit the pump cover.

5   Photographs showing removal and fitting of the pump are included in Chapter 1.

### 15  Oil pump: adjustment and bleeding of air

1   Refer to Routine Maintenance for details of pump adjustment
2   Bleed the pump of air whenever:
   a)   The machine has fallen on its side
   b)   The oil tank has run dry
   c)   Any part of the system has been disconnected

3   To bleed the pump and oil tank delivery pipe, first remove the pump cover. Check the oil tank is full. Hold a piece of clean, absorbent rag beneath the pump and loosen the bleed screw until oil appears. Wait until the oil running from the screw hole is free of air before retightening the screw.

4   To bleed the inlet stub feed pipe, start the engine and allow it to idle (below 2000 rpm). Set the pump stroke on maximum by pushing

fully up on the pump pulley. Observe the oil flow through the pipe and hold the pulley up until all air bubbles disappear. If the bubbles persist, check all pipe connections for security. On completion, stop the engine and replenish the oil tank. Check the pump cable for security and correct routing over the cable. Refit the pump cover.

15.3 Bleed the oil pump of air by loosening the bleed screw

## 16 Gearbox lubrication: general maintenance

1   Maintenance of the gearbox lubrication system consists of regular checking of the oil level and changing of the oil. Refer to Routine Maintenance for details of these operations.

## 17 Exhaust system: removal and refitting

1   Remove the exhaust pipe to cylinder retaining nuts. On AE models, remove the seat, fuel tank and right-hand sidepanel. Release the expansion chamber to silencer clamp, remove the chamber to frame retaining bolt and ease the chamber forward to clear the machine. Unbolt the silencer and ease it rearward to clear the machine. On AR models, remove the system to frame retaining bolts and remove the complete system.
2   On all models, note the fitted position of any mounting spacers. Renew flattened spring washers or damaged seals and mounting rubbers. Renew the seal between pipe and barrel.
3   Clean and examine the system and mounting components. Slight damage can be repaired by brazing a patch over the affected area. Refinish by wire brushing and degreasing the system and spraying with a suitable aerosol paint such as Sperex VHT.
4   When refitting, tighten each mounting finger-tight. Tighten the pipe to barrel nuts first, then the remaining mountings.
5   Treat any after-market system with caution, checking it is of reputable manufacture. Refer to the supplier for any alterations in carburettor jetting.

## 18 Exhaust system: decarbonisation

1   Refer to Routine Maintenance for details of this operation.

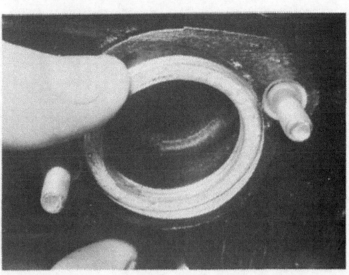

17.2 Always renew the exhaust pipe to cylinder barrel seal

# Chapter 3  Ignition system

## Contents

## Specifications

**Note:** The following specifications are for UK models only. Owners of US models check with the information at the end of this Section.

### Ignition system
Type ............................................................................................  CDI (capacitor discharge ignition) electronic

### Ignition timing (at 3000 rpm)
AE and AR50 ............................................................................  14°BTDC
AE80 A1 and AR80 A1 ..........................................................  17°BTDC
AE80 B1 and AR80 C1 ..........................................................  20°BTDC

### Ignition coil
Primary winding resistance ................................................  0.34 - 0.52 ohm
Secondary winding resistance ..........................................  3.2 - 4.8 K ohm

### Flywheel generator – except AE80 B1 and AR80 C1
Ignition source (exciter) coil resistance .........................  98 – 146 ohm
Pick-up coil resistance .......................................................  14 – 22 ohm

### Spark plug
Make .......................................................................................  NGK or ND
Type:
AE and AR50 ..........................................................................  B6ES or W20ES-U
AE80 A1 and AR80 A1 ..........................................................  BP8ES or W24EP-U
AE80 B1 and AR80 C1 ..........................................................  BPR8ES or W24EPR-U
Electrode gap .......................................................................  0.7 – 0.8 mm (0.027 – 0.031 in)

**US model specifications are as above except for the following information:-**

### Ignition timing (at 3000 rpm) ........................................  20°BTDC

### Flywheel generator
Ignition source (exciter) coil resistance .........................  66 – 86 ohm
Pick-up coil resistance .......................................................  16 – 22 ohm

### Spark plug
Make .......................................................................................  NGK or ND
Type:
AR50 ........................................................................................  B8ES or W24ES-U
AR80 ........................................................................................  BP7ES or W22EP-U

## 1 General description

Kawasaki AE and AR models feature what amounts to a maintenance-free electronic ignition system in which there are no moving parts and thus no mechanical wear. The CDI (capacitor discharge ignition) system comprises a solid state CDI unit and a magnetic pickup assembly. The latter replaces the traditional contact breaker assembly as the means of switching the ignition spark.

As the generator rotor moves, current is generated in the ignition source coil. This current is transferred to and stored in the capacitor in the CDI unit. As the generator rotor moves further, a trigger current is produced in the magnetic pick-up assembly which signals the stored power in the capacitor to discharge through the primary windings of the ignition coil. This surge of current induces a high voltage discharge from the ignition coil, which is fed via the HT lead to produce the spark across the electrodes of the spark plug.

**Warning:**

Take care to avoid electric shocks when checking the ignition system. The CDI unit produces a high voltage output which can prove very dangerous.

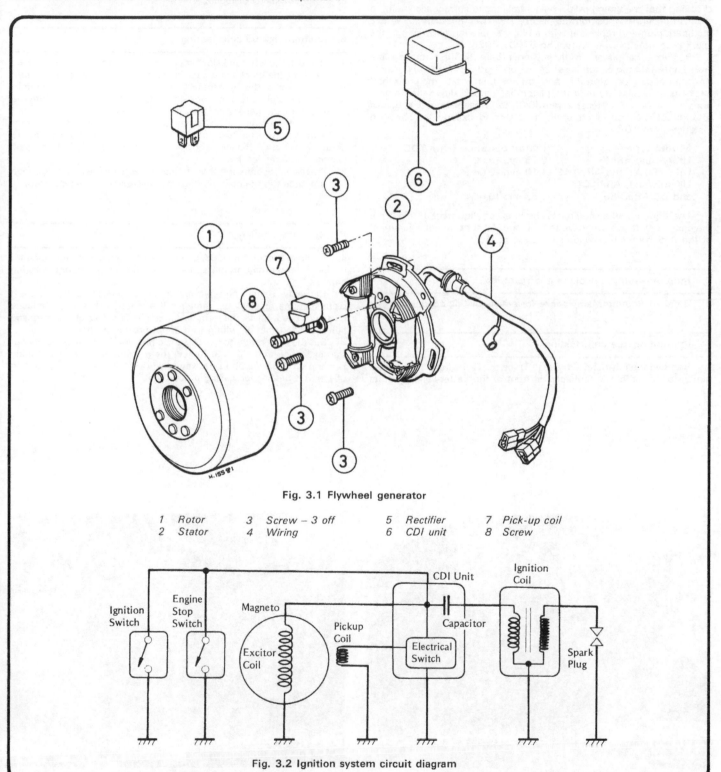

Fig. 3.1 Flywheel generator

| 1 | Rotor | 3 | Screw – 3 off | 5 | Rectifier | 7 | Pick-up coil |
|---|-------|---|---------------|---|-----------|---|--------------|
| 2 | Stator | 4 | Wiring | 6 | CDI unit | 8 | Screw |

Fig. 3.2 Ignition system circuit diagram

## 2 Ignition timing: verification of reference marks

1 The accuracy of timing depends very much on whether the generator rotor is set correctly on the crankshaft. Any wear between the dowel pin, crankshaft and rotor keyway will cause some amount of variation between the timing marks which will, in turn, lead to inaccurate timing. Inaccuracy in the timing mark position may also be a result of manufacturing error. The only means of overcoming this is to remove any movement between the two components and then to set the piston at a certain position within the cylinder bore before checking that the timing marks have remained in correct alignment. In order to accurately position the piston, it will be necessary to remove the spark plug and replace it with a dial gauge adapted to fit into the spark plug hole (Kawasaki tool no 57001-402).

2 Position the piston in the cylinder bore by first rotating the crankshaft until the piston is set in the top dead centre (TDC) position. Set the gauge at zero on that position and then rotate the crankshaft backwards (clockwise) until the piston has passed down the cylinder bore a distance of at least 2 mm (0.08 in). Reverse the direction of rotation of the crankshaft until the piston is exactly the specified distance from TDC.

| Model type | Piston distance from TDC |
| --- | --- |
| UK AE and AR50 | 0.75 mm (0.03 in) |
| UK AE80 A1 and AR80 A1 | 1.10 mm (0.04 in) |
| UK AE80 B1, AR80 C1 and US AR50/80 | 1.53 mm (0.06 in) |

The timing marks should now be in exact alignment. If this is not the case, new marks will have to be made. All subsequent adjustment of the timing may be made using these marks.

## 3 Ignition timing: checking and resetting

1 Refer to Routine Maintenance for details of this operation.

## 4 Ignition source coil: testing

1 Remove the left-hand sidepanel. Trace the leads from the flywheel generator to the block connector in front of the battery and separate the two halves of the connector. Check the terminals of the connector are free of moisture and corrosion and that the wiring from the connector to the generator is not broken or chafed.

2 Set a multimeter to its resistance function. Connect one meter probe to the black/red lead terminal of the connector and the other to earth. The meter reading should equal that specified if the windings of the coil are serviceable.

3 Kawasaki recommend that the coil is tested whilst cold. An open or short circuit in the coil windings will necessitate its replacement. stator renewal. Alternatively, consult an experienced auto-electrician as to the possibility of repair.

## 5 Ignition pick-up coil: testing

1 Remove the left-hand sidepanel. Trace the leads from the flywheel generator to the block connector in front of the battery and separate the two halves of the connector. Check the terminals of the connector are free of moisture and corrosion and that the wiring from the connector to the generator is not broken or chafed.

2 Set a multimeter to its resistance function. Connect one meter probe to the green/red lead terminal of the connector and the other to earth. The meter reading should equal that specified if the coil windings are serviceable.

3 Kawasaki recommend that the coil is tested whilst cold. An open or short circuit in the coil windings will necessitate its replacement.

## 6 CDI unit: testing

1 Remove the left-hand sidepanel. The CDI unit is mounted forward of the battery; slide it from its mounting and unplug it from the wiring harness.

2 Use only a small portable multimeter to test the unit. Kawasaki warn that the unit will be damaged if a meter with a large capacity battery is used. Check the terminals of the unit are free of moisture and corrosion and that all wiring to it is not broken or chafed.

3 Set the multimeter to its resistance function (K ohm) and carry out the tests shown in Fig. 3.3. Renew the unit if any one meter reading is incorrect. If in doubt, ask a Kawasaki dealer to verify the condition of the unit before purchasing a new item.

5.1 The ignition pick-up coil is mounted on the generator stator

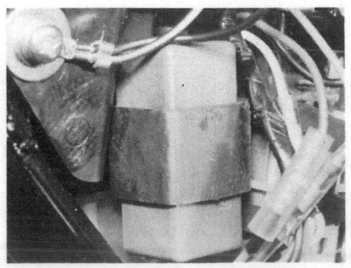

6.1 The CDI unit is contained within the battery compartment

| | Tester Positive (+) Lead Connection | | | | |
| --- | --- | --- | --- | --- | --- |
| Terminal | Pickup Coil | Ignition Coil | Ground | Ignition Switch | Exciter Coil |
| Pickup Coil | | ∞ | 15 − 50 kΩ (9 − 20 kΩ) | 200 kΩ − ∞ (15 − 40 kΩ) | 50 − 500 kΩ (15 − 40 kΩ) |
| Igntion Coil | ∞ | | ∞ | ∞ | ∞ |
| Ground | 5 − 20 kΩ (∞) | ∞ | | 2 − 6 kΩ (2 − 5 kΩ) | 2 − 6 kΩ (2 − 5 kΩ) |
| Igniiton Switch | ∞ | ∞ | ∞ | | 0 Ω * (0 − 0.5 Ω) |
| Exciter Coil | ∞ | ∞ | ∞ | 0 Ω * (0 − 0.5 Ω) | |

(Left header spanning rows: Tester Negative (−) Lead Connention)

Note: results shown in parentheses can be expected from alternative unit
*value also given in K ohms

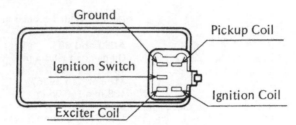

Fig. 3.3 CDI unit test

## 7 Ignition coil: testing

1 Remove the fuel tank. Pull the suppressor cap from the spark plug, disconnect it from the HT lead, then disconnect the coil low tension (LT) lead.
2 Set a multimeter to its resistance function (K ohm). Connect one meter probe to the end of the HT lead and the other to earth. The meter reading should equal 3.2 − 4.8 K ohm if the secondary windings of the coil are serviceable.
3 Reset the meter to its ohms scale. Connect one probe to the LT terminal and the other to earth. The meter reading should equal 0.34 − 0.52 ohm if the primary windings of the coil are serviceable.
4 Kawasaki recommend that these tests be carried out with the coil cold. An open or short circuit in the coil windings will necessitate coil renewal. If in doubt, have a Kawasaki dealer test the coil on a spark gap tester.

## 8 High tension lead: examination

1 Erratic running faults can sometimes be attributed to leakage from the HT lead and spark plug. With this fault, it will often be possible to see tiny sparks around the lead and suppressor cap at night. One cause is dampness and the accumulation of road salts around the lead. It is often possible to cure the problem by drying and cleaning the components and spraying them with an aerosol ignition sealer.
2 If the system has become swamped with water, use a water dispersant spray. Renew the cap seals if defective. If the lead or cap is suspected of breaking down internally, renew the component.
3 Where the lead is permanently attached to the ignition coil, entrust its renewal to an auto-electrician who will have the expertise to solder on a new lead without damaging the coil windings.

## 9 Spark plug: checking and resetting the gap

1 Refer to Routine Maintenance for details of this operation.

7.1 The ignition coil is mounted beneath the fuel tank

# Chapter 4 Frame and forks

*Refer to Chapter 7 for information relating to the 1984 on models*

## Contents

## Specifications

**Note:** The following specifications apply to both UK and US models

### Frame

| | |
|---|---|
| Type ............................................ | Cradle, welded tubular steel |

### Front forks

| | AE50 and 80 | AR50 and 80 |
|---|---|---|
| Type ................................. | Oil damped telescopic | Oil damped telescopic |
| Travel .............................. | 140 mm (5.5 in) | 130 mm (5.1 in) |
| Spring free length ........... | 416 mm (16.38 in) | 358 mm (14.09 in) |
| Service limit .................... | 408 mm (16.06 in) | 351 mm (13.82 in) |
| Oil capacity (per leg) ...... | 92 ± 4 cc (3.24 ± 0.14 Imp fl oz) | 87 ± 4 cc (3.06 ± 0.14/2.94 ± 0.13 Imp/US fl oz) |
| Oil level* ......................... | 489.5 ± 2mm (19.27 ± 0.08 in) | 431 ± 2mm (16.97 ± 0.08 in) |
| Oil type ........................... | SAE 5W/20 | SAE 5W/20 |

*Fork oil level is measured from the top of the stanchion to the top of the oil, forks fully extended and top plugs, spring retaining plugs, spacer tubes, washers and springs removed*

### Rear suspension

**All models**

| | |
|---|---|
| Type ............................................ | Kawasaki Uni-Trak – Swinging arm subframe controlled by single suspension unit through linkage |
| Travel (at wheel spindle) ............... | 120 mm (4.7 in) |
| Pivot shaft maximum runout .......... | 0.7 mm (0.03 in) |
| Bush bore diameter: | |
| Rocker arm ................................. | 17.022 - 17.097 mm (0.670 - 0.673 in) |
| Service limit ............................... | 17.250 mm (0.679 in) |
| Swinging arm ............................. | 17.050- 17.100 mm (0.671 - 0.673 in) |
| Service limit ............................... | 17.250 mm (0.679 in) |
| Tie rod ....................................... | 17.022 - 17.247 mm (0.670 - 0.679 in) |
| Service limit ............................... | 17.350 mm (0.683 in) |
| Sleeve outside diameter ................. | 16.982 - 17.0 mm (0.668 - 0.669 in) |
| Service limit .................................... | 16.95 mm (0.667 in) |

### Torque wrench settings – kgf m (lbf ft)

| Component | AE50 and 80 | AR50 and 80 |
|---|---|---|
| Handlebar clamp bolts ........................................ | 2.1 (15.0) | N/App |
| Front fork spring retaining plugs ........................ | 2.0 (14.5) | N/App |
| Front fork top bolts ........................................... | N/App | 3.0 (22.0) |
| Front fork damper rod retaining bolts ................. | 2.3 (16.5) | 1.8 (13.0) |
| Upper yoke pinch bolts ...................................... | 1.0 (7.2) | N/App |
| Lower yoke pinch bolts ....................................... | 2.1 (15.0) | 1.8 (13.0) |

| Component | All models |
|---|---|
| Steering stem top bolt ........................................ | 2.0 (14.5) |
| Steering head bearing adjuster ring ..................... | 2.0 (14.5) |
| Rear suspension unit securing nuts ..................... | 3.3 (24.0) |
| Rear suspension tie rod and rocker arm securing nuts .......... | 3.3 (24.0) |
| Swinging arm pivot shaft nut .............................. | 4.0 (29.0) |

## 1 General description

The frame is of conventional welded tubular steel construction, having a single front downtube which divides into a duplex cradle at its base.

The front forks are the conventional telescopic type, having internal oil-filled damping.

Rear suspension is Kawasaki Uni-Trak, incorporating a fabricated subframe which pivots on the main frame whilst being controlled by a single oil-filled suspension unit operating through a linkage.

## 2 Front fork legs: removal and refitting

1 Refer to Chapter 5 and remove the front wheel.

### AE models

2 Slacken each fork yoke pinch bolt and pull each leg downwards to clear the yokes, having first removed its plastic top plug.

### AR models

3 Remove the front mudguard and brace. Note the fitted position of spacers and washers and renew flattened spring washers.

4 When removing the left-hand leg, detach the two brake caliper securing bolts, free the caliper from the leg and tie it to a point on the frame. Ensure the wedge placed between the brake pads during wheel removal remains in position.

5 Remove each fork top bolt. Release each fairing stay to fork leg clamp. If removing both legs together, remove the fairing. Slacken the lower yoke pinch bolts and pull each leg downwards to clear the yoke.

### All models

6 Reverse the removal procedure when fitting each fork leg. On AE models, align the top of each stanchion with the top surface of the upper yoke before fully tightening the yoke pinch bolts. On AR models, tighten the fork top bolt before the lower yoke pinch bolt. On all models, tighten the wheel spindle nut before the yoke pinch bolts and note the specified torque wrench settings.

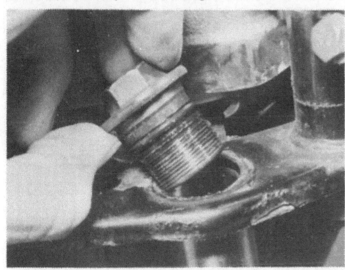

2.5a On AR models, remove each fork leg top bolt ...

2.5b ... and release the fairing before removing the leg

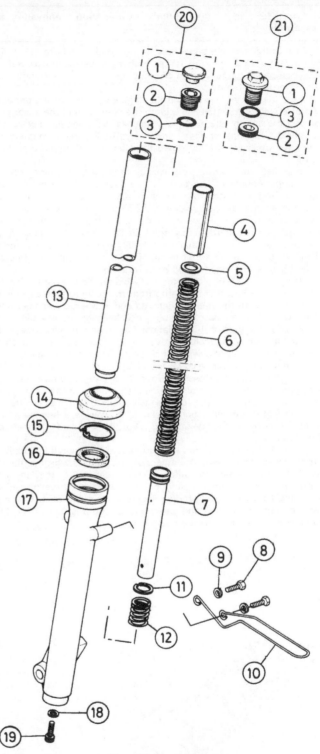

Fig. 4.1 Front forks

| | | | |
|---|---|---|---|
| 1 | Top plug | 11 | Damper rod piston ring |
| 2 | Spring retaining plug | 12 | Rebound spring |
| 3 | O-ring | 13 | Stanchion |
| 4 | Spacer tube | 14 | Dust excluder |
| 5 | Washer | 15 | Circlip |
| 6 | Spring | 16 | Oil seal |
| 7 | Damper rod | 17 | Lower leg |
| 8 | Bolt – 2 off | 18 | Sealing washer |
| 9 | Spring washer – 2 off | 19 | Damper rod retaining bolt |
| 10 | Cable guide – AE model only | 20 | AE models |
| | | 21 | AR models |

## 3   Front fork legs: dismantling, examination, renovation and reassembly

1   Dismantle each leg separately, using an identical procedure. This will prevent parts being unwittingly exchanged between legs. Refer to the accompanying figure when dismantling and lay out the component parts in order of removal.

2   With the leg vertical, clamp its stanchion between the protected jaws of a vice. Alternatively, reclamp the stanchion in the lower yoke. Remove the spring retaining plug. Release the stanchion and remove the spacer tube, washer and spring. Invert the leg over a container and pump the lower leg up and down the stanchion to assist draining of the oil.

3   It is likely the damper rod will rotate with its retaining bolt thus preventing removal of the bolt. To manufacture a rod holding tool, obtain a length of steel bar which is approximately 0.5 inch thick, square in section and approximately 20 inches long. Cut one end of this bar to a point, as shown in the figure accompanying this text. It is now necessary to harden this point so that when it is pushed into position against the circular recess in the end of the damper rod, its edges will bite into the softer metal of the recess walls thus providing enough grip to prevent the rod from rotating.

4   Using the flame of a blowlamp or welding torch, heat the bar to a cherry red for about half its length from the pointed end. Directly the bar turns red with the heat, quench its end in a large container full of water. Only 1-2 inches of the bar need be submerged beneath the surface of the water for this initial cooling procedure. Once the end of the bar has cooled, quickly remove it from the water and polish the edges of its pointed end with emery cloth. The heat remaining in the uncooled section of bar will soon travel, by conduction, to the tip of the point. As this happens, the spectrum of tempering colours will be seen to progress up the edges of the point. Upon seeing the straw colour appear on the point edges, quench the complete tool in the water until it is completely cooled. The bar should now have a hardened point with the metal gradually decreasing in hardness towards the mid point of its length.

5   When hardening the bar, take great care to observe the following safety precautions. Always be aware of the dangers which come from naked flames and heated metal; have a means of extinguishing fire nearby and wear both eye and body protection. Thrusting red hot metal into cold water will produce a very violent reaction between the two; be prepared for this and guard against the possibility of scalding water being thrown from the container.

6   Clamp a self-grip wrench to the bar end or drill a hole through which to pass a tommy bar. Grip the fork lower leg horizontally between the protected jaws of a vice. Locate the point of the special tool in the damper rod recess, giving the tool a sharp tap to seat it. Unscrew the damper rod retaining bolt and withdraw the tool and damper rod. Several attempts may be needed before the tool successfully grips the rod.

7   Remove the dust excluder and pull the stanchion out of the lower leg. Remove the oil seal from the lower leg only if it is to be renewed. Displace the seal retaining clip with the flat of a small screwdriver and then use a larger screwdriver to lever out the seal whilst taking care not to damage the alloy edge of its housing.

8   Measure the spring length and compare it with that specified or with a new spring. If a spring has taken a permanent set to a shorter length, renew both springs as a set. Check the spring spacer and washer for wear and renew if necessary.

9   Renew the oil seal and O-ring(s) as a matter of course. Renew the dust excluder if split or perished. It is advisable to protect the exposed stanchions by fitting gaiters instead of the standard excluders.

10   Renew the piston ring of the damper rod if damping has deteriorated. If in doubt about spring condition, then compare it with a new item. Check the rod oilways are free from obstruction.

11   Check the stanchion for straightness by rolling it on a flat surface. A bent stanchion or one with peeling chrome must be renewed. Scuffing and penetration of the chromed surface will indicate wear and the need for renewal of the stanchion and possibly the lower leg.

12   Thoroughly clean each component part and place it on a clean work surface. Lightly lubricate the oil seal and O-rings with clean fork oil.

13   Reassembly is a reversal of the dismantling procedure. Use a length of metal tube to drift the oil seal into position. The tube end making contact with the seal must be square to the lower leg, properly chamfered and free of burrs.

14   Before fitting the damper rod retaining bolt, check its sealing washer is serviceable and coat both of its sides with sealing compound. Coat the threads of the bolt with locking compound and tighten it to the specified torque loading.

15   Before fitting the spring retaining plug, replenish the leg with the correct quantity of specified oil. On AE models, tighten the plug to the specified torque loading. On AR models, refer to the accompanying figure and fit the plug so that its top surface is 22 mm (0.86 in) from the top of the stanchion.

16   Check each leg for correct operation before fitting it to the machine.

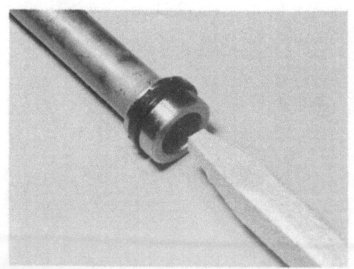

3.3 The damper rod must be held by a special tool

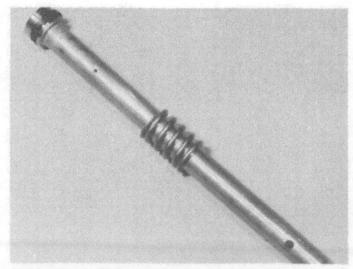

3.10 Examine the damper rod assembly

3.13 Retain the lower leg oil seal with the wire clip

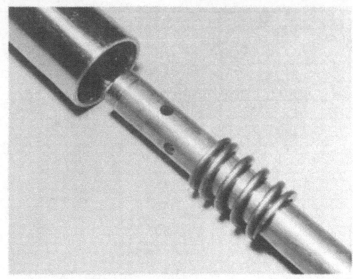

3.14a Slide the damper rod into the stanchion ...

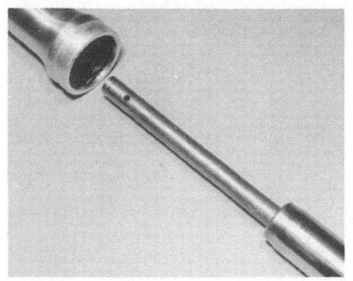

3.14b ... slide the stanchion into the lower leg ...

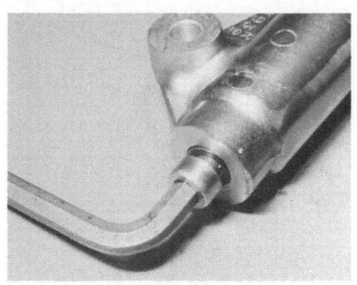

3.14c ... fit the damper rod retaining bolt and sealing washer

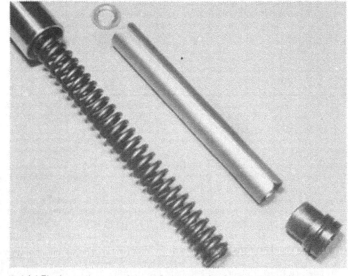

3.14d Fit the spring, washer and spacer tube into the stanchion ...

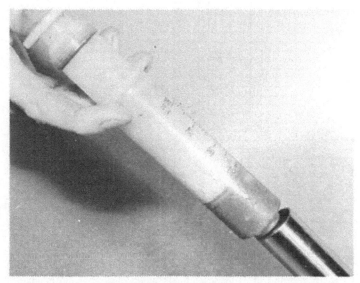

3.15a ... before fitting the spring retaining plug, replenish the fork leg with oil

3.15b Fit serviceable dust excluders or better still, fork leg gaiters

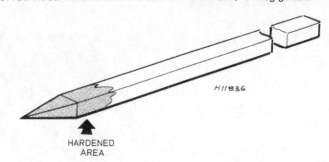

HARDENED
AREA

**Fig. 4.2 Fabricated tool for holding the damper rod in position**

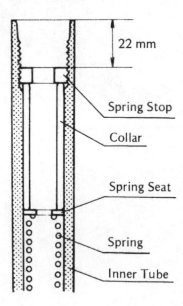

22 mm

Spring Stop

Collar

Spring Seat

Spring

Inner Tube

**Fig. 4.3 Refitting the front fork spring retaining plug – AR models**

## 4  Steering head assembly: removal and refitting

1   Refer to Chapter 5 and remove the front wheel.
2   Refer to Section 2 and remove the front fork legs. On AE models, unbolt the mudguard from the lower yoke.
3   Isolate the battery by disconnecting one of its leads; this will prevent shorting of exposed contacts when disconnecting headstock components.

4   Use a commonsense approach when detaching components from the headstock. Remove only those components necessary to permit detachment of the upper and lower yokes.
5   Note the fitted position of the handlebars, marking them if necessary. Protect the tank with rag, detach the bars and rest them on it. If necessary, detach the left-hand switch and its wires from the bars. Detach the headlamp and instrument console mounting bracket from the yokes (4 bolts) and ease it forward to clear the headstock.
6   Remove the steering stem top bolt and washer. Remove the upper yoke, tapping it upwards with a soft-faced hammer to free it.
7   Support the lower yoke. Using a C-spanner, remove the bearing adjuster ring. Make provision to catch any balls that may fall from the bearings, remove the dust excluder and cone of the upper bearing and lower the yoke and stem from position.
8   Refer to the following Section for bearing examination and renovation.
9   When refitting the assembly, retain the balls around the lower cone with grease of the recommended type. Fill both bearing cups with grease. Tighten the adjuster ring finger-tight. Refer to Routine Maintenance and carry out bearing adjustment. Note the specified torque wrench settings.
10  On completion, check all electrical leads and control cables are correctly routed. Check the headlamp beam height, see Chapter 6. Check all controls and instruments are functioning correctly.

## 5  Steering head bearings: examination and renovation

1   The bearing cup and cone tracks should be free from indentations or cracks. If necessary, renew the cups and cones as complete sets.
2   If any one ball is cracked or blemished, renew the complete set. Do not risk refitting defective balls, they are relatively cheap. There are twenty three No. 6 ($3/16$ in) balls in both the upper and lower races; this leaves a gap between any two balls which must not be filled otherwise wear will occur.
3   Drift each bearing cup from the steering head if necessary, using a long drift passed through the opposite end of the head. Ensure the cup leaves its location squarely by moving the drift around it.
4   The lower cone can be drifted off the steering stem with a flat chisel, keeping it square to the stem.
5   Before fitting new cups, clean and lightly grease their locations. Support the steering head on a wooden block whilst drifting each cup home. Using a short length of metal tube of the same outer diameter as the cup, with a piece of wood placed across it, will provide the most effective method of driving the cup squarely home. The tube ends must be square to its length. Clean and grease the stem before fitting the lower cone.

## 6  Fork yokes: examination

1   Renew any yoke which is damaged or cracked. Check for distortion by pushing the fork stanchions through the lower yoke and fitting the top yoke. If alignment is correct, the yokes are not bent.

## 7  Steering lock: renewal

1   Once operated, the lock prevents the handlebars from turning once they are set on full lock. If the lock malfunctions it must be renewed, repair is impracticable.
2   Some dismantling of the steering head may be necessary to allow removal of the lock securing screws. Obtain a key with the new lock and carry it when riding the machine.
3   Cases have occurred of the lock retaining screws slackening in use, which could result in the lock being lost. To prevent this, examine the screws at regular intervals to ensure that they are tight; there should be a dab of paint across the head of each to permit an easy visual check of whether the screws have slackened off. If no paint is found remove the lock and thoroughly clean the screw threads. Use a degreasing agent to finish off, allow them to dry, then refit the lock applying a few drops of Loctite 270 or similar thread-locking compound to the screw threads. Tighten the screws securely and evenly, then apply a dab of paint across the head of each.

## 8 Speedometer and tachometer: examination and renovation

1   If either instrument fails to function, check its drive cable for breakage. A dry cable or one which is trapped or kinked will cause the instrument needle to jerk. Refer to Routine Maintenance for cable care. Refer to Chapter 5 for details of the speedometer drive gearbox and to Chapter 1 for details of the tachometer drive (AR models only).

2   It is not practicable to repair an instrument head; obtain a serviceable item. A speedometer in correct working order is required by law.

3   To remove each instrument head, detach the headlamp from its bracket and ease it forward. Disconnect the drive cable by unscrewing its retaining ring. Unplug the bulb holders from the base of the instrument console and unscrew the two console to mounting bracket retaining nuts. Pull the console clear of the bracket, separate its two halves and detach the instrument head from the lower half by removing the two screws at its base.

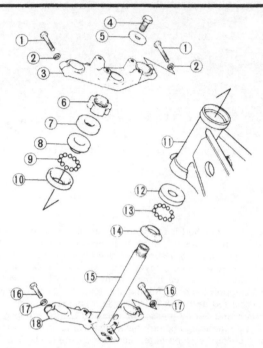

Fig. 4.4 Steering head assembly – AE models

| 1 | Bolt – 2 off | 10 | Upper bearing cup |
|---|---|---|---|
| 2 | Spring washer – 2 off | 11 | Frame |
| 3 | Upper yoke | 12 | Lower bearing cup |
| 4 | Top bolt | 13 | Lower bearings |
| 5 | Washer | 14 | Lower bearing cone |
| 6 | Bearing adjuster ring | 15 | Steering stem |
| 7 | Dust excluder | 16 | Bolt – 2 off |
| 8 | Upper bearing cone | 17 | Spring washer – 2 off |
| 9 | Upper bearings | 18 | Lower yoke |

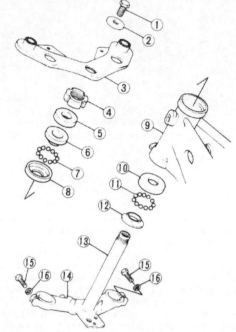

Fig. 4.5 Steering head assembly – AR models

| 1 | Top bolt | 9 | Frame |
|---|---|---|---|
| 2 | Washer | 10 | Lower bearing cup |
| 3 | Upper yoke | 11 | Lower bearings |
| 4 | Bearing adjuster ring | 12 | Lower bearing cone |
| 5 | Dust excluder | 13 | Steering stem |
| 6 | Upper bearing cone | 14 | Lower yoke |
| 7 | Upper bearings | 15 | Bolt – 2 off |
| 8 | Upper bearing cup | 16 | Spring washer – 2 off |

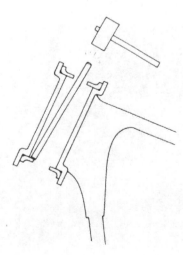

Fig. 4.6 Method of removing steering head bearing cups

8.3a Release each instrument drive cable ...

8.3b ... separate the two halves on the instrument console ...

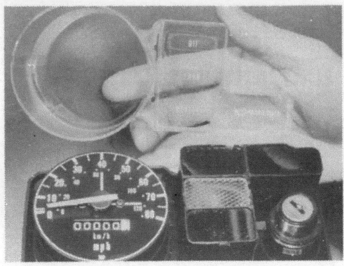

8.3c ... and remove the instrument shield (AE shown)

## 9  Swinging arm: removal, examination, renovation and refitting

1    Wear in the arm pivot will cause imprecise handling with a tendency for the rear of the machine to twitch or hop. Support the machine with its rear wheel clear of the ground. Hold onto the main frame and pull and push the rear of the arm. Play in the pivot will be magnified by the leverage effect produced.

2    Wear will necessitate renewal of the pivot bushes. Refer to Chapter 5 and remove the rear wheel. Move the drive chain clear of the arm. Detach the arm from the suspension linkage and main frame by removing the necessary bolts. Manoeuvre the arm rearwards and down to clear the machine.

3    Thoroughly clean the arm and pivot components. Examine the arm for distortion and cracking. If repair is expensive, look around for a good secondhand item. Remove corrosion with a wire brush and derusting agents and replace the paint finish.

4    Renew the pivot shaft if its shank is stepped, badly scored or bent beyond the specified limit. No wear should exist between the shaft, sleeve and pivot bushes. Refer to Specifications for wear limits.

5    Do not attempt to remove the pivot bushes unless they are to be renewed; they will be destroyed upon removal. Make sure new bushes are obtainable. Drift each bush from the arm by using a long drift passed through the opposite end of the housing. Ensure the housing is well supported and each bush leaves it squarely.

6    Clean and lightly grease the housing before fitting new bushes. Support the housing and drift each bush home, using a short length of metal tube of the same outer diameter as the bush. Check that tube ends are square to its length.

7    Renew the chain deflector fitted over the left-hand end of the bush housing, if perished or worn. Renew any flattened spring washers. Grease and fit the pivot sleeve.

8    Align the arm in the frame. Lightly grease and fit the pivot shaft, checking each washer is correctly positioned. Tighten the shaft nut to the specified torque loading. Check that the arm pivots smoothly before reconnecting the suspension linkage.

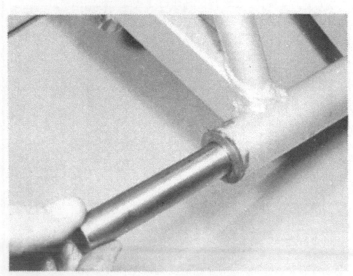

9.4 Check the swinging arm pivot sleeve for wear ...

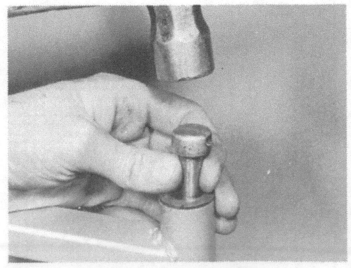

9.5 ... remove the pivot bushes only if unserviceable

9.7 The chain deflector must be in good condition

**Fig. 4.7 Swinging arm and Uni-Trak assembly**

| | | | |
|---|---|---|---|
| 1 | Nut – 5 off | 16 | Tie-rod |
| 2 | Spring washer – 5 off | 17 | Bush |
| 3 | Washer – 9 off | 18 | Pivot shaft |
| 4 | Suspension unit | 19 | Spring washer |
| 5 | Bush – 3 off | 20 | Nut |
| 6 | Rocker arm | 21 | Pivot bolt |
| 7 | Sleeve – 2 off | 22 | R-pin |
| 8 | Bush – 3 off | 23 | Nut |
| 9 | Bolt – 5 off | 24 | Spring washer |
| 10 | Sleeve – 2 off | 25 | Brake torque arm |
| 11 | Washer | 26 | Bolt |
| 12 | Chain deflector – AE model | 27 | Chain adjuster – 2 off |
| 13 | Pivot shaft bush – 2 off | 28 | Washer – 2 off |
| 14 | Swinging arm | 29 | Adjusting nut – 2 off |
| 15 | Bush | 30 | Locknut – 2 off |

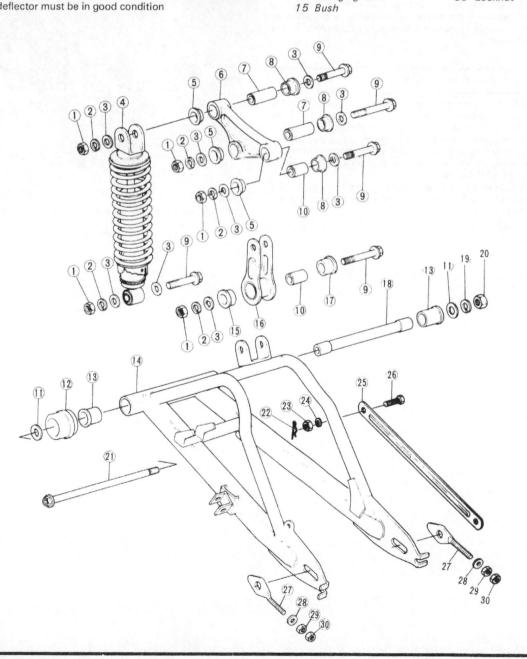

## 10 Uni-Trak suspension: removal, examination, renovation and refitting

1   Refer to the preceding Section and remove the swinging arm. When removing the suspension unit and linkage, work in a logical sequence and place each component on the work surface in the order of removal.

2   The suspension unit is sealed. Leakage, failure of damping, damage to the piston housing, a bent or corroded damper rod or deterioration of its rubber mounting bushes will necessitate unit renewal. Thoroughly clean the unit but do not attempt to grease the damper rod, this will cause seal failure.

3   Observe the necessary fire precautions and clean each linkage component with a petrol/paraffin mix, drying it thoroughly on completion. Check for play between each pivot bolt and its sleeve, bushes and housing. Note the wear limits given in Specifications. Excessive wear, scoring or signs of overheating (blueing) will necessitate immediate renewal of the component concerned. Obtain specialist advice if the frame and swinging arm lugs are worn. Renew any flattened spring washers.

4   When assembling and refitting the suspension components, take care not to introduce dirt into the pivots. Grease each pivot component. Check that the arrow on the rocker arm is pointing forwards. Check all washers are positioned correctly and tighten all securing nuts to the specified torque settings. Once assembled, check each pivot moves freely.

## 11 Frame: examination and renovation

1   If the frame is bent through accident damage, replacement is the most satisfactory solution; look around for a good secondhand item. Rejigging is possible but is a specialist task and can be expensive.

2   Clean the frame regularly, checking the welded joints for cracking. Minor damage can be repaired by welding or brazing. Remove corrosion with a wire brush and derusting agents, it can cause reduction in the material thickness.

3   A misaligned frame will cause handling problems and may promote 'speed wobbles'. If necessary, strip the machine completely and have the frame checked.

## 12 Footrests, prop stand and brake pedal: general maintenance

1   Frequently examine and lubricate each component; failure or incorrect operation can cause a serious accident. Check for free movement over the full operating range. Renew worn pivot components and fatigued or badly corroded springs. Check for security; split-pins must be correctly fitted.

2   If a component is bent, remove it and clamp it in a vice. Using a blow lamp or welding torch, heat the area around the bend to a cherry red. Carefully straighten the component by tapping it with a hammer.

3   Straightening a component with the metal cold will cause stress fractures, leading to fatigue. Before refitting the component, check for signs of failure and renew if in doubt.

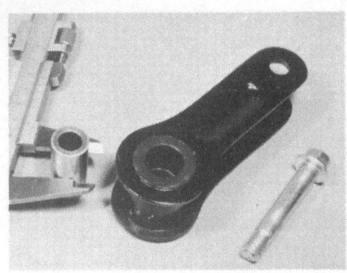

10.3 Examine each pivot assembly of the Uni-Trak suspension

# Chapter 5 Wheels, brakes and tyres

*Refer to Chapter 7 for information relating to the 1984 on models*

## Contents

## Specifications

**Note**: The following specifications apply to both UK and US models

### Front disc brake

| | |
|---|---|
| Type ........................................................... | Single hydraulic disc |
| Fluid specification ..................................... | SAE J1703 (UK) or DOT 3 (US) |
| Pad lining thickness ................................... | 3.7 mm (0.145 in) |
| Service limit .............................................. | 1.0 mm (0.04 in) |
| Disc thickness ........................................... | 3.8 – 4.1 mm (0.149 – 0.161 in) |
| Service limit .............................................. | 3.5 mm (0.138 in) |
| Disc runout (maximum) ............................... | 0.3 mm (0.012 in) |
| Master cylinder: | |
|    Bore diameter ..................................... | 11.0 – 11.063 mm (0.433 – 0.435 in) |
|    Service limit ........................................ | 11.08 mm (0.436 in) |
|    Piston diameter .................................... | 10.823 – 10.850 mm (0.426 – 0.427 in) |
|    Service limit ........................................ | 10.80 mm (0.425 in) |
|    Primary cup diameter ........................... | 11.3 – 11.7 mm (0.445 – 0.461 in) |
|    Service limit ........................................ | 11.2 mm (0.441 in) |
|    Spring free length ............................... | 38.3 – 42.3 mm (1.508 – 1.665 in) |
|    Service limit ........................................ | 36.4 mm (1.433 in) |
| Caliper: | |
|    Bore diameter ..................................... | 30.23 – 30.28 mm (1.190 – 1.192 in) |
|    Service limit ........................................ | 30.30 mm (1.193 in) |
|    Piston diameter .................................... | 30.167 – 30.200 mm (1.187 – 1.189 in) |
|    Service limit ........................................ | 30.13 mm (1.186 in) |

### Front and rear drum brakes

| | |
|---|---|
| Type ........................................................... | Internally expanding, single leading shoe, drum |
| Shoe lining thickness: | |
|    UK .................................................... | 3.2 – 3.9 mm (0.126 – 0.153 in) |
|    US .................................................... | 2.9 – 3.5 mm (0.114 – 0.138 in) |
| Service limit: | |
|    UK .................................................... | 1.8 mm (0.071 in) |
|    US .................................................... | 1.6 mm (0.063 in) |
| Drum internal diameter ............................... | 110.0 – 110.087 mm (4.331 – 4.334 in) |
| Service limit .............................................. | 110.75 mm (4.360 in) |
| Cam shaft diameter .................................... | 11.957 – 11.984 mm (0.471 – 0.472 in) |
| Service limit .............................................. | 11.88 mm (0.467 in) |
| Backplate cam shaft hole diameter ............... | 12.0 – 12.027 mm (0.472 – 0.473 in) |
| Service limit .............................................. | 12.15 mm (0.478 in) |
| Shoe spring free length ............................... | 30.8 – 31.2 mm (1.212 – 1.228 in) |
| Service limit .............................................. | 32.6 mm (1.283 in) |

## Final drive chain

Type .................................................................................... EK420G

No of links:

AE and AR50 ............................................................ 116

AE and AR80 ............................................................ 114

Play ...................................................................... 30 – 35 mm (1.2 – 1.4 in)

| | AE50 and 80 | AR50 and 80 |
|---|---|---|
| **Wheels** | | |
| Type ...................................... | Conventional, steel-spoked with chromed steel rim | Cast aluminium alloy |
| Size: | | |
| Front ........................... | 1.40 x 19 | 1.40 x 18 |
| Rear ............................. | 1.60 x 16 | 1.40 x 18 |
| Rim runout (maximum): | | |
| Axial .......................... | 2.0 mm (0.08 in) | 0.5 mm (0.02 in) |
| Radial ........................ | 2.0 mm (0.08 in) | 0.8 mm (0.03 in) |
| **Tyres** | | |
| Size: | | |
| Front ........................... | 2.50 – 19 4PR | 2.50 – 18 4PR |
| Rear ............................. | 3.00 – 16 4PR | 2.75 – 18 4PR (UK) |
| | | 2.75 – 18 6PR (US) |
| Pressures: | | |
| Front ........................... | 1.5 kg/cm² (21 psi) | 1.75 kg/cm² (25 psi) |
| Rear: | | |
| UK ........................... | 2.25 kg/cm² (32 psi) | 2.25 kg/cm² (32 psi) |
| US ........................... | N/App | Up to 97.5 kg (215 lb) load – 2.0 kg/cm² (28 psi) 97.5 – 180 kg (215 – 397 lb) load – 2.8 kg/cm² (40 psi) |

*Note: loads given represent total weight of rider, passenger and any accessories or luggage. Pressures apply to original equipment tyres only – if different tyres are fitted check with tyre manufacturer or supplier whether different pressures are necessary*

| | | |
|---|---|---|
| Manufacturer's recommended minimum tread depth ...................... | 2.0 mm (0.08 in) | Front – 1.0 mm (0.04 in) Rear – 2.0 mm (0.08 in) |

## Torque wrench settings – kgf m (lbf ft)

| Component | | |
|---|---|---|
| Front wheel spindle nut ........................................................ | 3.3 (24.0) | 6.5 (47.0) |
| Rear wheel spindle nut ......................................................... | 6.0 (43.0) | 6.5 (47.0) |
| Rear torque arm nuts: | | |
| To frame ........................................................... | 3.3 (24.0) | 3.3 (24.0) |
| To brake backplate .......................................... | 3.0 (22.0) | 3.0 (22.0) |
| Sprocket nuts or bolts ......................................................... | 2.1 (15.0) | 2.1 (15.0) |
| Brake disc bolts .................................................................... | N/App | 2.1 (15.0) |
| Master cylinder clamp bolts ................................................. | N/App | 0.9 (6.5) |
| Handlebar lever pivot locknut .............................................. | N/App | 0.6 (4.3) |
| Caliper mounting bolts ......................................................... | N/App | 3.3 (24.0) |
| Hydraulic hose union bolts ................................................... | N/App | 3.0 (2.0) |

## 1  General description

### AE models

Each wheel comprises a chromed steel rim laced to an aluminium alloy hub by steel spokes. The tubed tyres have a block pattern tread suitable for both on and off-road use. Each brake is drum with single leading shoe operation.

### AR models

Each five-spoke cast alloy wheel is fitted with a conventional tubed tyre. The hydraulically-operated, single disc, front brake utilises a single piston caliper. The mechanically-operated rear brake is drum with single leading shoe operation.

## 2  Wheels: examination and renovation

1  Refer to Routine Maintenance for details of this operation.

## 3  Front wheel: removal and refitting

1  Position a stout wooden crate or block beneath the engine so that the machine is well supported, with the front wheel clear of the

ground. Remove the wheel spindle retaining nut and washer (where fitted).

### AE models

2  Detach the speedometer cable from the brake backplate by unscrewing its knurled retaining ring and pulling it clear. Remove the nut from the brake cable end and pull the cable clear of the backplate. Retain the nut, trunnion, spring and gaiter to prevent loss. Renew the spring if fatigued or broken. Renew the gaiter if split or perished.

3  Support the wheel and pull the spindle clear. If necessary, use a soft-metal drift and hammer to tap it from position. Remove the wheel, noting the fitted position of washer and spacer.

4  Refer to the accompanying figure and check the spacer is correctly fitted before lifting the wheel into position. Ensure the fork leg spigot engages in the backplate slot otherwise the wheel will lock directly the brake is applied, with disastrous consequences.

5  Grease the spindle before insertion and tap it lightly to seat it. Tighten the retaining nut to the specified torque loading.

6  Reconnect the speedometer and brake cables. Refer to Routine Maintenance and adjust the brake. Recheck all disturbed connections for security before taking the machine on the road.

### AR models

7  Support the wheel and pull the spindle clear. Pull the wheel forward to clear the brake caliper, detach the speedometer gear box

from the hub and remove the wheel, noting the fitted position of the spacer.

8 When leaving the wheel removed, slip a strip of wood between the brake pads to stop them being expelled if the lever is accidentally squeezed. Check the spacer collar remains located in the centre of the speedometer drive gear, retain it with grease if necessary.

9 When fitting the speedometer gear box to the hub, ensure the tangs of the drive plate engage correctly in the hub slots. Avoid damage to the brake pads by carefully sliding the disc between them. Check the spacer is correctly fitted between wheel and fork leg. The spigot of the speedometer gear box must engage in the fork leg slot, see accompanying photograph.

10 Grease the spindle before insertion and tap it lightly to seat it. Tighten the retaining nut to the specified torque loading. Refer to Routine Maintenance and check brake operation and fluid level. Recheck all disturbed connections for security before taking the machine on the road.

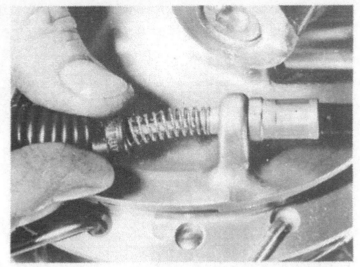

3.2 On AE models, examine the front brake cable gaiter and return spring ...

3.4 ... ensure the fork leg spigot engages in the backplate slot ...

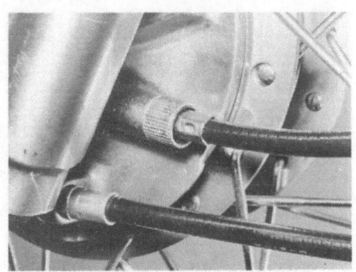

3.6 ... and reconnect the brake and speedometer cables

3.7 On AR models, fit the front wheel spacer ...

3.8 ... position the spacer collar in the speedometer drive gear ...

3.9 ... and engage the gear casing spigot in the fork leg slot

**Fig. 5.1 Front wheel – AE models**

1  Left-hand bearing
2  Spacer
3  Wheel hub
4  Right-hand bearing
5  Oil seal
6  Spacer
7  Washer
8  Nut
9  Bolt
10  Cam operating arm
11  Wear indicator plate
12  O-ring
13  Brake backplate
14  Cam shaft
15  Drive plate
16  Oil seal
17  Return spring – 2 off
18  Brake shoe – 2 off
19  Speedometer drive gear
20  Nut
21  Pin
22  Washer
23  Speedometer driven gear
24  Cable housing
25  Wheel spindle

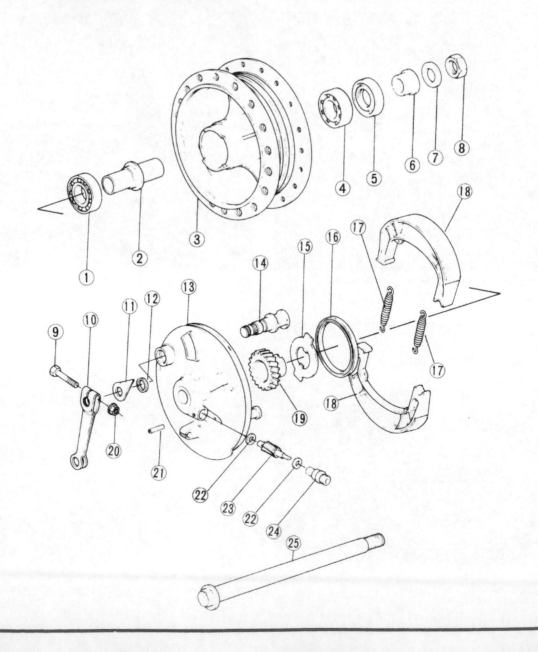

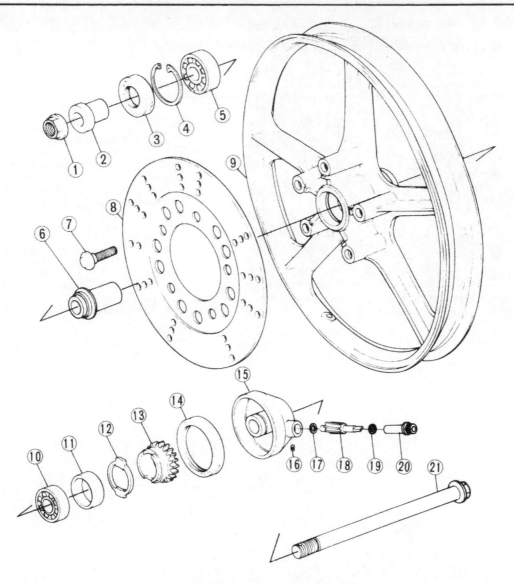

Fig. 5.2 Front wheel – AR models

| 1 | Nut | 7 | Bolt – 5 off | 13 | Speedometer drive gear | 17 | Washer |
|---|-----|---|--------------|----|------------------------|----|--------|
| 2 | Spacer | 8 | Brake disc | 14 | Oil seal | 18 | Speedometer driven gear |
| 3 | Oil seal | 9 | Front wheel | 15 | Speedometer gearbox | 19 | Washer |
| 4 | Circlip | 10 | Right-hand bearing | | housing | 20 | Cable housing |
| 5 | Left-hand bearing | 11 | Spacer collar | 16 | Grub screw | 21 | Wheel spindle |
| 6 | Spacer | 12 | Drive plate | | | | |

## 4  Rear wheel: removal and refitting

1   Position a stout wooden crate or blocks beneath the engine so that the machine is well supported with the rear wheel clear of the ground.

2   Remove the wheel spindle retaining nut. Remove the spring-pin, nut and spring washer which retain the torque arm to the brake backplate. Detach the arm. Remove the split-pin (where fitted) and nut from the brake rod end. Depress the brake pedal so the rod leaves the trunnion of the operating arm. Retain the nut, trunnion and spring to prevent loss. Renew the spring if fatigued or broken. Renew the split-pin.

3   Lay a length of clean rag beneath the final drive chain. Rotate the wheel to align the chain split link with the wheel sprocket. Using flat-nose pliers, remove the spring clip from the link. Remove the link, lift the chain ends off the sprocket and place them on the rag.

4   Support the wheel and pull out the spindle. If necessary, use a soft-metal drift and hammer to tap it from position. Remove the wheel, noting the fitted position of each spacer.

5   Refer to the accompanying figure and check each spacer is correctly fitted before lifting the wheel into position. Grease the spindle before insertion and tap it lightly to seat it. Fit the retaining nut, finger-tight.

6   Reconnect the brake rod and torque arm. Fit the new split-pin through the rod (where applicable) and bend its legs apart. Renew the arm spring washer, if flattened, fit the nut finger-tight and fit a serviceable spring clip.

7   Reconnect and adjust the chain, see Routine Maintenance. Note instructions for fitting the split link. Tighten the spindle retaining nut and torque arm nut to the specified torque settings.

8   Refer to Routine Maintenance and adjust the brake. Recheck all disturbed connections for security before taking the machine on the road.

4.6 Reconnect the rear brake rod and torque arm, using a serviceable spring clip through the arm bolt (AE shown)

**Fig. 5.3 Rear wheel – AE models**

| | | | |
|---|---|---|---|
| 1 | Nut | 14 | Nut – 4 off |
| 2 | Spacer | 15 | Tab washer – 2 off |
| 3 | Oil seal | 16 | Return spring – 2 off |
| 4 | Left-hand bearing | 17 | Brake shoe – 2 off |
| 5 | Sprocket | 18 | Brake backplate |
| 6 | Final drive chain | 19 | Bolt |
| 7 | Bolt – 4 off | 20 | Spacer |
| 8 | Spring clip | 21 | Wheel spindle |
| 9 | Plate | 22 | Cam shaft |
| 10 | Split link | 23 | O-ring |
| 11 | Wheel hub | 24 | Wear indicator plate |
| 12 | Spacer | 25 | Cam operating arm |
| 13 | Right-hand bearing | 26 | Nut |

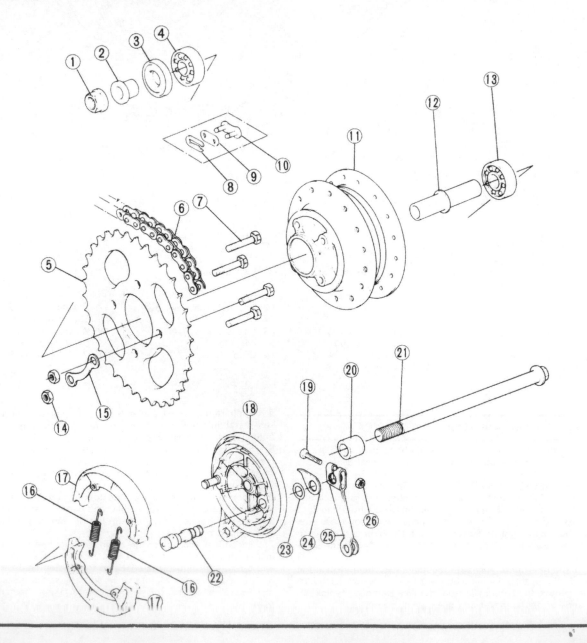

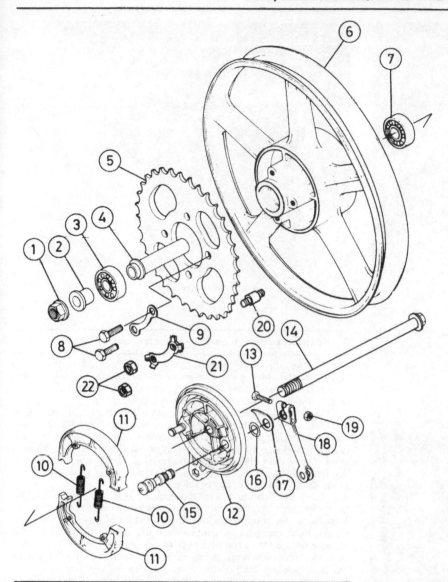

**Fig. 5.4 Rear wheel – AR models**

1 Nut
2 Spacer
3 Left-hand bearing
4 Spacer
5 Sprocket
6 Rear wheel
7 Right-hand bearing
8 Bolt – 4 off ▲
9 Tab washer – 2 off ▲
10 Return spring – 2 off
11 Brake shoe – 2 off
12 Brake backplate
13 Bolt
14 Wheel spindle
15 Cam shaft
16 O-ring
17 Wear indicator plate
18 Cam operating arm
19 Nut
20 Stud – 4 off △
21 Tab washer – 2 off △
22 Nut – 4 off △

▲ early models
△ later models

## 5 Wheel bearings: removal, examination and refitting

1   Remove the wheel. Where a drum brake is fitted, detach the brake backplate. Where a disc brake is fitted, remove the brake disc (see Section 11). The disc will not prevent bearing removal, Kawasaki recommend its removal to avoid distortion. Note the bearing retaining circlip fitted to the front wheel of AR models. Using a screwdriver, carefully lever the oil seal from position and remove the circlip.

2   Support the wheel hub on wooden blocks placed as close to the bearing as possible. Place the end of a long drift against the upper face of the lower bearing, moving the centre spacer sideways. Tap the bearing downwards, moving the drift around the bearing so that it leaves the hub squarely.

3   Any oil seal will be removed with the bearing and should be renewed as a matter of course. Withdraw the spacer, invert the wheel and drift out the second bearing.

4   Use petrol to clean all grease from the hub and bearings. Observe the necessary fire precautions. Spin each bearing and check for play or roughness. Renew if in doubt.

5   Before fitting, pack each bearing with the recommended grease. Clean and lightly grease each bearing housing. Check the housings for abnormal wear caused by movement of the bearing outer race. Obtain specialist advice if a bearing is a loose fit in the hub.

6   Where applicable, fit the bearing with its sealed side facing outboard. Select a socket which has an overall diameter slightly less than the bearing outer race. Support the hub and use the socket and a soft-faced hammer to drift the first bearing into its housing. Keep the bearing square to the hub otherwise the housing surface may be broached.

7   Invert the wheel and fit the spacer. The spacer must be correctly positioned, see accompanying figure. Fit the second bearing. On AR models, fit the bearing retaining circlip, checking it locates firmly in its groove. Renew the circlip if loose or distorted. Use the socket and hammer to tap home each new oil seal. Lightly grease the seal lip.

5.2 Drift each wheel bearing from the hub

5.7a Fit the first bearing followed by the spacer

5.7b Use a socket to tap home each bearing

5.7c Lightly grease the lip of each bearing oil seal

## 6  Speedometer drive gear: examination and renovation

1    Remove the front wheel. On AE models, detach the brake backplate; the drive gear is fitted in the backplate. On AR models, detach the cable from the gear box by unscrewing its knurled retaining ring. Remove the spacer collar from the gear.

2    On all models, remove the drive plate and gear. Clean each part and the worm gear in the housing. Do not remove the worm gear unless damaged; this is difficult to achieve. On AR models, a small grub screw holds the gear retainer in place. On AE models, a pin is used. Rather than risk damaging the housing, obtain specialist advice.

3    Renew any part which is obviously worn. Badly worn drive plate tangs are the normal cause of drive failure.

4    If defective, carefully lever the large oil seal which surrounds the gear from position. Use the flat of a large screwdriver and take care not to damage the housing. On AE models, failure of this seal will allow grease from the gear to contaminate the brake linings. Clean the housing before fitting the new seal. Check the seal enters its location squarely. Start with finger pressure then place a strip of wood across the seal, tapping down on the wood to drive the seal home.

5    Lightly grease each part of the drive before fitting. Check the drive plate turns smoothly and transmits drive to the worm gear. Grease the lip of the oil seal.

6.3a Remove the speedometer drive gear seal if unserviceable ...

6.3b ... and examine the drive assembly (AE shown)

**Tyre changing sequence - tubed tyres**

 **A** Deflate tyre. After pushing tyre beads away from rim flanges push tyre bead into well of rim at point opposite valve. Insert tyre lever adjacent to valve and work bead over edge of rim.

 **B** Use two levers to work bead over edge of rim. Note use of rim protectors

**C** Remove inner tube from tyre

**D** When first bead is clear, remove tyre as shown

 **E** When fitting, partially inflate inner tube and insert in tyre

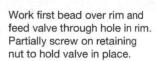

**F** Work first bead over rim and feed valve through hole in rim. Partially screw on retaining nut to hold valve in place.

 **G** Check that inner tube is positioned correctly and work second bead over rim using tyre levers. Start at a point opposite valve.

**H** Work final area of bead over rim whilst pushing valve inwards to ensure that inner tube is not trapped

### 7 Wheel sprocket: examination and renewal

1 Refer to Routine Maintenance for details of this operation.

### 8 Final drive chain: examination, adjustment and lubrication

1 Refer to Routine Maintenance for details of these operations.

### 9 Brakes: adjustment and wear check

1 Refer to Routine Maintenance for details of these operations.

### 10 Drum brakes: dismantling, examination, renovation and reassembly

1 It is false economy to cut corners with brake components; the safety of machine and rider depends on their good condition.

2 Remove the wheel and pull the brake backplate from the hub. If the brake linings are contaminated by oil or grease, or they have worn beyond the specified limit, renew the shoes.

3 Remove surface dirt with a petrol-soaked rag. The dirt contains asbestos and is harmful if inhaled. Ease down high spots with a file.

There is no satisfactory method of degreasing the linings.

4 If necessary, remove the shoes by folding them together into a 'V' and pulling them off the backplate. Detach the springs and examine them for fatigue or failure. Refer to Specifications and measure the spring free length; renew the springs if overstretched.

5 Remove the cam operating arm, marking its fitted position. Remove the wear indicator, O-ring and cam shaft. Renew the O-ring, if flattened or perished. Clean the cam shaft and its location in the brake backplate. Renew the shaft or backplate if worn beyond the specified limit. Severe scoring of the mating surfaces will also necessitate renewal. Grease and fit the shaft, positioning it correctly. Fit the O-ring and position the wear indicator so that it points to the extreme right of the scale on the backplate. Fit the cam operating arm. Lubricate the cam face and the pivot stud with brake grease.

6 Before refitting existing shoes, break the surface glaze with glasspaper. Do not inhale any dust. Assemble the springs and shoes and reverse the removal procedure to fit them to the backplate. Do not risk distortion by using excessive force.

7 Examine the drum surface for scoring, oil contamination or wear beyond the service limit, all of which will impair braking efficiency. Remove dust with a petrol-soaked rag; the dust is harmful if inhaled. Use a rag soaked in petrol to remove oil deposits; observe the necessary fire precautions.

8 If the drum is deeply scored, it must be skimmed on a lathe or renewed. Excessive skimming will adversely affect brake performance; consult a specialist.

10.2 Remove the drum brake shoes from the backplate ...

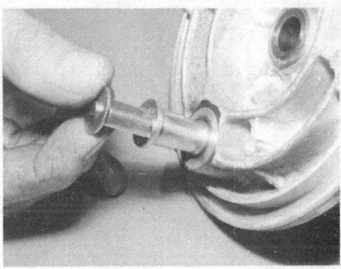

10.5a ... examine the cam shaft ...

10.5b ... fit the O-ring and position the wear indicator as shown ...

10.5c ... refit the cam operating arm

## 11 Disc brake: dismantling, examination, renovation and reassembly

### Brake disc

1   Check the disc for wear and distortion with the wheel fitted. Using a micrometer, measure the disc thickness at the point of greatest wear and renew the disc if the measurement is less than 3.5 mm (0.138 in).

2   Set up a suitable pointer close to the outer periphery of the disc. Spin the wheel slowly and measure the amount of distortion (run out). Renew the disc if the measurement is more than 0.3 mm (0.012 in). A badly distorted disc will cause juddering during braking, the brake to bind when not in use and reduce braking efficiency.

3   Renew the disc if its contact area with the pads is badly scored. To renew it, first remove the wheel. Remove the five disc retaining bolts and ease the disc away from the hub. Do not use excessive force.

4   Clean the disc to hub mating surfaces before fitting the new disc. Check the surfaces are together and tighten the retaining bolts evenly and in a diagonal sequence to the specified torque loading.

### Hydraulic hose

5   Check the hose for signs of leakage, damage, deterioration or scuffing against cycle components. Check the end unions are tight and in good condition with no stripped threads or damaged sealing washers. Do not tighten the union bolts over the specified torque setting.

6   In the interests of safety, Kawasaki recommend hose renewal every four years.

### Brake pads

7   Refer to Routine Maintenance for details of brake pad examination and renewal

### Caliper

8   Detach the hydraulic hose from the caliper and allow the brake fluid to drain into a suitable container. Avoid spillage; brake fluid is an efficient paint stripper and will damage plastic components. Wipe up any spillage immediately.

9   Remove the two caliper to fork leg mounting bolts and ease the caliper clear of the disc. Refer to Routine Maintenance and displace the brake pads. Note the fitted position of the caliper mounting bracket and pull it clear of the caliper. Remove the two dust seals. Do not blow away brake dust, it is harmful if inhaled.

10   Remove the piston head. Carefully use the flat of a screwdriver to displace the piston dust seal. Place a thick wad of rag over the end of the caliper bore and apply a jet of compressed air to the hose union orifice to displace the piston. Protect the eyes from any blowback of fluid which may escape the rag. Do not use too high an air pressure.

11   Clean the caliper components thoroughly, only in brake fluid. Never use petrol or cleaning solvent. In the interests of safety, renew all rubber components as a matter of course.

12   Check the piston and caliper bore for scoring, rust or pitting all of which will prevent a good fluid seal. Renew each part if affected or worn beyond the specified service limit.

13   Clean the spindles of the mounting bracket and lubricate them sparingly with Poly Butyl Cuprysil grease (PBC is a high temperature, water resistant grease). Remove the bleed screw and check it for blockage. Renew the screw sealing cap if damaged or missing. Refit the bleed screw.

14   During assembly, maintain ultra-clean conditions. Dirt particles will score bearing surfaces and cause early failure. Check the piston seal is not twisted in its retaining groove. Lubricate the seal, piston and bore with new brake fluid before fitting. Check the piston dust seal is properly located in the bore and piston grooves before fitting the piston head.

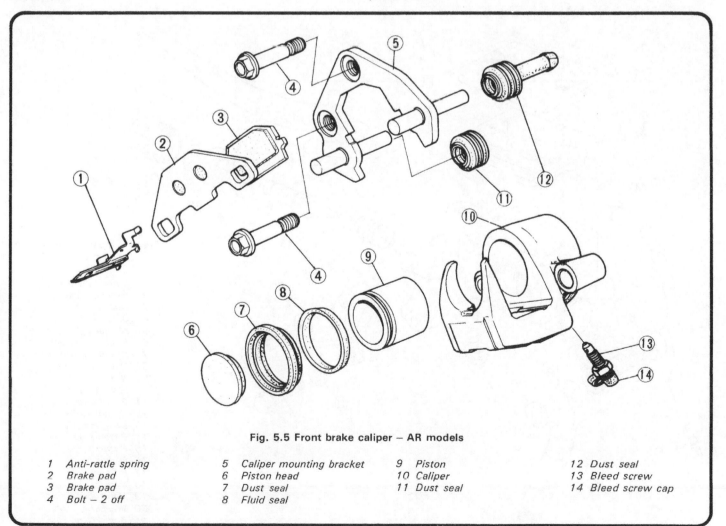

**Fig. 5.5 Front brake caliper — AR models**

| 1 | Anti-rattle spring | 5 | Caliper mounting bracket | 9 | Piston | 12 | Dust seal |
|---|---|---|---|---|---|---|---|
| 2 | Brake pad | 6 | Piston head | 10 | Caliper | 13 | Bleed screw |
| 3 | Brake pad | 7 | Dust seal | 11 | Dust seal | 14 | Bleed screw cap |
| 4 | Bolt – 2 off | 8 | Fluid seal | | | | |

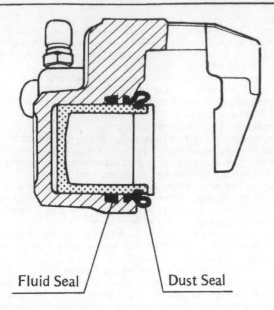

Fluid Seal          Dust Seal

**Fig. 5.6 Piston seal locations**

15 Slide the mounting bracket into the caliper. Refer to Routine Maintenance and fit the anti-rattle spring and pads. Carefully slide the caliper over the disc. Tighten the mounting bolts to the specified torque loading.
16 Renew the hose union washers, if damaged. Tighten the union bolt to the specified torque loading. Refill the master cylinder with new brake fluid. Refer to Routine Maintenance and bleed the system. Check the brake for correct operation several times, check for leakage and take the machine for a test run using the brake as often as possible. Recheck for fluid loss.

**Master cylinder**
17 Attach a plastic tube to the caliper bleed screw and run it into a suitable container. Open the screw and operate the brake lever to empty the system of fluid. Retighten the screw. Remove the handlebar fairing. Unscrew the rear view mirror.
18 Detach the hydraulic hose from the cylinder, placing rag beneath the union to catch any fluid. Brake fluid will strip paint and damage plastic components; wipe up any spillage immediately.
19 Remove the brake lever from the cylinder. Trace the leads from the stop lamp switch to their connectors in the headlamp shell. Disconnect the leads and pull them free.
20 Support the cylinder so that it remains upright and remove its clamp. Carefully lift the cylinder away from the machine. Remove the reservoir cap and diaphragm and empty any fluid within the unit into a container.
21 Use the flat of a small screwdriver to release the locating tabs of the cylinder liner. Withdraw the dust seal, piston stop, piston, primary

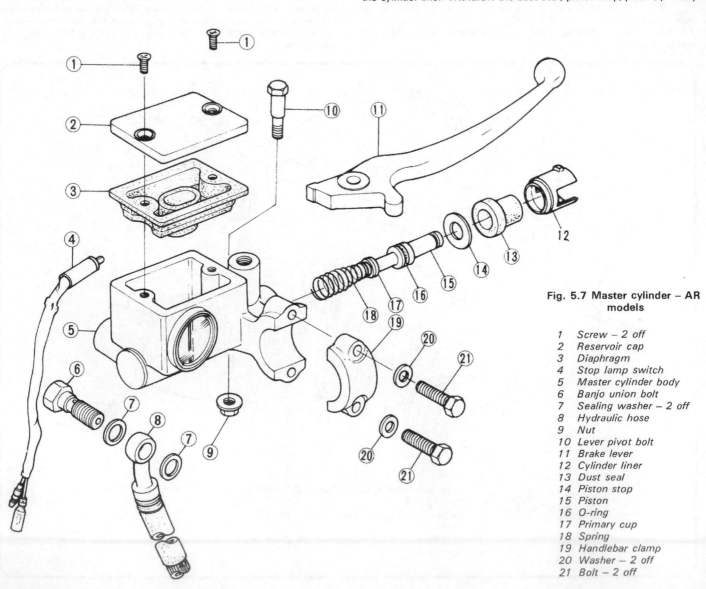

**Fig. 5.7 Master cylinder – AR models**

1   Screw – 2 off
2   Reservoir cap
3   Diaphragm
4   Stop lamp switch
5   Master cylinder body
6   Banjo union bolt
7   Sealing washer – 2 off
8   Hydraulic hose
9   Nut
10  Lever pivot bolt
11  Brake lever
12  Cylinder liner
13  Dust seal
14  Piston stop
15  Piston
16  O-ring
17  Primary cup
18  Spring
19  Handlebar clamp
20  Washer – 2 off
21  Bolt – 2 off

cup and spring. Clean the cylinder components thoroughly, only in clean brake fluid. Never use petrol or cleaning solvent.

22 Inspect the cylinder body for signs of stress failure around the lever pivot, handlebar mounting, hose union boss and mirror mounting in particular. Renew the body if cracked or if the cylinder bore is scored, pitted or worn beyond the specified limit.

23 Renew the piston O-ring if flattened or damaged. Measure the primary cup diameter, compare with the specified wear limit and renew if overworn. Note the fitted position of the cup. Renew the dust seal and liner, if damaged.

24 Measure the piston diameter and renew if less than the specified wear limit. Renew the spring, if broken, or set to less than the free length limit. Renew the diaphragm if split or perished.

25 Maintain ultra-clean conditions during assembly. Dirt particles will score bearing surfaces and cause early failure. Lubricate the piston assembly with new brake fluid before fitting. Note the specified torque settings for the clamp bolts, lever pivot locknut and hose union bolt. Renew the hose union washers, if damaged.

26 Refill the master cylinder with new brake fluid. Refer to Routine Maintenance and bleed the system. Check the brake for correct operation several times, check for leakage and take the machine for a test run using the brake as often as possible. Recheck for fluid loss. Check the stop lamp for correct operation.

**Bleeding the hydraulic system**

27 Refer to Routine Maintenance for details of this operation.

---

## 12 Valve cores and caps

1 Dirt under the valve seat will cause a puzzling 'slow-puncture'. Check for leaks by applying spittle to the valve and watching for bubbles.

2 The cap is a safety device and should always be fitted. It keeps dirt out of the valve and provides a second seal in case of valve failure, thus preventing an accident resulting from sudden deflation.

---

## 13 Tyres: removal, repair and refitting

1 Remove the wheel and deflate the tyre by removing the valve core. Push the tyre bead on both sides away from the wheel rim and into its centre well. Remove the locking ring and push the valve into the tyre.

2 Insert a tyre lever close to the valve and lever the tyre bead over the outside of the rim. On AR models, protect the cast alloy rim by fitting suitable protectors fabricated from short lengths of thick-walled nylon fuel pipe split down one side with a sharp knife. Work around the rim until one side of the tyre is completely free. No great force is required; if resistance is encountered check both beads have fully entered the rim well. Remove the inner tube.

3 Work from the other side of the wheel and ease the other bead over the outside of the rim furthest away until the tyre is free.

4 If the tube is punctured, inflate it and immerse in water. Bubbles will indicate the source of the leak which should be marked. Deflate and dry the tube. Clean the punctured area with a petrol soaked rag. When dry, apply rubber solution. With the solution dry, remove the patch backing and apply the patch.

5 Self vulcanising patches are the best type, It may be necessary to remove a protective covering from the top of the patch after application. Synthetic rubber tubes may require a special patch and adhesive.

6 If the tube is already patched or if it is torn, renew it. Sudden deflation can cause an accident.

7 Before fitting the tyre, check its inside and outside for the cause of the puncture. Do not fit a tyre with damaged tread or sidewalls. On AE models the rim tape must be fitted; this prevents the spoke ends chafing the tube and causing punctures. Fitting can be aided by dusting the tyre beads with french chalk. Washing up liquid can be used but may cause corrosion of the inner rim, use it sparingly to avoid tyre creep.

8 Inflate the tube just enough for it to assume a circular shape, no more, and push it fully into the tyre. Lay the tyre on the rim at an angle. Insert the valve through the rim hole and screw the locking ring on the first few threads.

9 Starting at the point furthest from the valve, push the tyre bead over the rim and into the centre well. Work around the tyre until the whole of one side is on the rim. If necessary, use a tyre lever during the final stage.

10 Check there is no pull on the valve and repeat the procedure to fit the second bead into the rim. Finish adjacent to the valve, pushing it into the tyre as far as the locking ring will allow, thus ensuring the tube is not trapped when the last section of bead is levered over the rim.

11 Check the tube is not trapped; reinflate it to the specified pressure and tighten the valve locking ring. Check the tyre is correctly seated on the rim. A thin rib moulded around each tyre wall should be equidistant from the rim at all points. If the tyre is uneven on the rim, try bouncing the wheel to reseat it. Fit the valve dust cap.

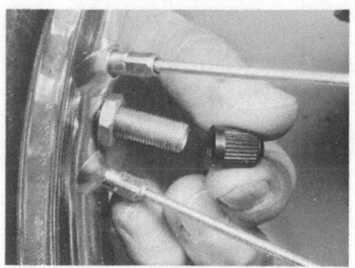

12.2 The valve cap is a safety device and should always be fitted

# Chapter 6 Electrical system

*Refer to Chapter 7 for information relating to the 1984 on models*

## Contents

## Specifications

**Note**: The following specifications are for UK models only. Owners of US models check with the information at the end of this Section.

### Battery
| | |
|---|---|
| Make | Yuasa, GS or FB |
| Type | 6N6 – 1C |
| Voltage | 6 volt |
| Capacity | 6 Ah |
| Electrolyte specific gravity | 1.26 at 20° (68°F) |
| Earth | Negative |

### Fuse
| | |
|---|---|
| Rating | 20 amp |

### Resistor
| | |
|---|---|
| Rating | 25W, 2 ohm |

### Flywheel generator

Rated output at 8000 rpm:

| | | |
|---|---|---|
| AE50, AR50, AE80 A1, AR80 A1 | 1.0A/8.5V | |
| AE80 B1, AR80 C1 | 2.7A/8.8V | |
| Lighting circuit (AC) voltage | 6.5V approx at 4000 rpm | |

| Charging circuit output at 4000 rpm: | Lights off (day) | Lights on (night) |
|---|---|---|
| AE50, AR50, AE80 A1, AR80 A1 | 0.8 – 1.0A/8.0 – 8.5V | 0.6 – 1.2A/7.0 – 8.0V |
| AE80 B1, AR80 C1 | N/Av | 2.0A/8.8V |

Generator coil resistance (cold) – AE50, AR50, AE80 A1, AR80 A1:

| | |
|---|---|
| Charging coil – Pink wire to earth | 0.28 – 0.42 ohm |
| Lighting coil – Yellow wire to earth | 0.25 – 0.37 ohm |

### Bulbs
| | |
|---|---|
| Headlamp | 6V, 25/25W |
| Pilot lamp | 6V, 3W |
| Tail/stop lamp | 6V, 5/21W |
| Direction indicator lamps | 6V, 21W |
| Instrument illuminating light | 6V, 3W |
| Neutral indicator light | 6V, 3W |
| Direction indicator warning light | 6V, 1.7W |
| Oil level warning light | 6V, 3W |

**US model specifications are as shown except for the following information:-**

### Resistor
| | |
|---|---|
| Resistor | Not fitted – lights permanently on |

### Flywheel generator
| | |
|---|---|
| Rated output | 1.8A/8.8V at 8000 rpm |

Charging circuit output at 4000 rpm:

| | |
|---|---|
| Lights off (headlamp Black/yellow wire disconnected) | 2.5A/8.5V |
| Lights on | 1.5A/8.5V |

Generator coil resistance (cold):

| | |
|---|---|
| Charging coil – Pink wire to earth | 0.26 – 0.38 ohm |
| Lighting coil – Yellow wire to earth | 0.16 – 0.24 ohm |

## Bulbs

| | |
|---|---|
| Headlamp ............................................................... | 6V, 30/30W – sealed beam |
| Tail/stop lamp ........................................................ | 6V, 5.3/25W (3/32 cp) |
| Direction indicator lamps .................................... | 6V, 17W (21 cp) |
| Main beam warning light ..................................... | 6V, 1.7W |

## 1 General description

The flywheel generator stator incorporates a coil to provide lighting and battery charging power and one to provide ignition source power.

The charging coil produces alternating current which is converted to direct current by a silicone diode rectifier to make it compatible with the battery and system components. The rectifier effectively blocks half of the output wave by acting as a one-way electronic switch; this system is known as half-wave rectification. Lighting is provided by a tap taken off the charging coil which feeds alternating current direct to the main lighting circuit. When the lights are not in use, a ballast resistor absorbs the same amount of power, thereby obviating any risk of the lighting coil current combining with that of the charging coil to overcharge the battery.

In the event of a short circuit or sudden surge, the circuit is protected by a 10 amp fuse incorporated in the battery positive lead. The fuse forms a weak link which will blow and thus prevent damage to the circuit and components.

## 2 Testing the electrical system

1   Continuity checks, used when testing switches, wiring and connections, can be carried out using a battery and bulb arrangement to provide a test circuit. For most tests however, a pocket multimeter should be considered essential.

2   A basic multimeter capable of measuring volts and ohms can be purchased for a reasonable sum and will prove invaluable. Separate volt and ohm meters can be used, provided those with correct operating ranges are available. If generator output is to be checked, an ammeter of 0-4 amperes range will be required.

3   Take care when performing any electrical test, some electrical components can be damaged if incorrectly connected or inadvertently earthed. Note instructions regarding meter probe connections.

4   If in doubt, or where test equipment is not available, seek professional assistance. Do not risk damaging expensive electrical parts.

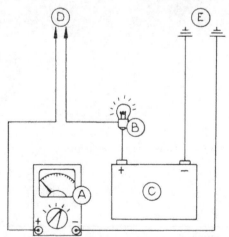

**Fig. 6.1 Simple testing arrangement for checking the electrical system**

| | | |
|---|---|---|
| A   Multimeter | C   Battery | E   Negative probe |
| B   Bulb | D   Positive probe | |

## 3 Wiring: layout and examination

1   The wiring is colour-coded, corresponding with the accompanying

wiring diagram. Socket connectors are designed so that reconnection can only be made correctly.

2   Check for breaks or frayed outer coverings which will cause short circuits. A wire may become trapped, breaking the inner core but leaving the resilient outer cover thereby causing mysterious intermittent or total circuit failure. Corroded or badly made connections will also cause problems.

3   Intermittent short circuits can often be traced to a chafed wire which passes through or close to a metal component. Avoid situations where a wire is tightly bent or can be trapped between moving parts.

## 4 Lighting/charging coil: testing

### Charging coil performance

1   Remove the left-hand side panel to expose the battery leads. *Without disconnecting the battery*, connect a voltmeter between them (meter positive + terminal to battery positive terminal +). Set the meter to its 0 – 10 volt DC range and check that the lights are switched off (on US models, disconnect the headlamp Black/yellow wire). Start the engine and increase its speed to 4000 rpm. Note the meter reading and stop the engine. The meter reading should equal 8.0 – 8.5 volt if the coil is serviceable.

2   Disconnect the white/red lead of the rectifier from the battery positive (+) lead. Connect the positive probe of an ammeter to the white/red lead. Connect the meter negative probe to the battery positive lead. Check all connections are clean and secure. Set the ammeter to its dc range. Start the engine and increase its speed to 4000 rpm. With the lights off, note the meter reading and stop the engine. The meter reading should equal that specified if the coil is serviceable.

### Lighting coil performance

3   Detach the headlamp rim, complete with reflector and glass, from the shell. Set a voltmeter to its 0 – 10 ac range and connect its positive probe to the red/yellow lead from the lighting switch. Earth the meter negative probe. Switch the lighting to 'On' and select main beam. Start the engine and increase its speed to 4000 rpm. Check all lighting is on, note the meter reading and stop the engine. The meter reading should equal 6.5 volt if the coil is serviceable.

### Resistance checks

4   If the results of the above tests prove unsatisfactory continue by making resistance tests of the coils. Refer to the accompanying wiring diagram and identify the lead running from the coil in question by the colour coding. Disconnect the generator from the main wiring loom.

5   Set a multimeter to its resistance function and measure for resistance across the coil windings. When connecting one meter probe to earth, select a clean, unprotected point on the crankcase. The meter reading should correspond with that specified.

6   If the check is unsatisfactory, the coil is unserviceable and must be renewed. If it transpires that the coils are good, but poor charging or lighting performance is still experienced, continue with the tests in the next two Sections after having determined the condition of the rotor magnets.

## 5 Rectifier: testing

1   The rectifier must be kept clean and dry and mounted so that it is not exposed to direct contamination by oil or water yet has free circulation of air to permit cooling. It can be damaged by inadvertently reversing the battery connections.

2   To test, set a multimeter to its resistance function (x1 ohm range) and connect its probes to the unit connectors. Note the meter reading and reverse the probes. If one reading shows continuity and the other non-continuity, the unit is serviceable.

## 6 Ballast resistor: testing (UK models)

1   If overcharging is experienced, indicated by a rapid level drop of the battery electrolyte, or if frequent bulb blowing occurs there is evidence that the resistor has failed.

2   To test the resistor, set a multimeter to its resistance function, disconnect the two wires leading to the resistor and place the probes of the meter on the wire terminals to measure the resistance across the resistor. If the reading obtained differs markedly from the specified resistance of 2 ohm, the resistor is unserviceable and must be renewed.

## 7 Battery: examination and maintenance

1   Normal maintenance requires keeping the electrolyte level between the upper and lower marks on the battery case and checking the vent tube is correctly routed and not blocked. Unless acid is spilt, top up with distilled water.

2   If electrolyte level drops rapidly, suspect over-charging or leakage. A cracked battery case cannot be effectively repaired and will necessitate battery renewal. Acid spillage must be neutralised with an alkali (washing soda or baking powder) and washed away with fresh water, otherwise serious corrosion will occur. Top up with sulphuric acid of 1.260 specific gravity.

3   Warpage of the plates and separators indicates an expiring battery as does sediment filling the gap between the case bottom and plates.

4   Check the lead connections are tight and free from corrosion. If necessary, remove corrosion by scraping with a knife, finishing with emery cloth. Remake the connections and smear with petroleum jelly (not grease) to prevent further corrosion.

5   Recharging is required when the acid specific gravity falls below 1.260 (at 20°C - 60°F). Take the reading at the top of the meniscus with the hydrometer vertical. It is advisable to give the battery a 'refresher' charge every six weeks or so if the machine is rarely used. If left discharged for too long, the battery plates will sulphate and inhibit recharging. Refer to the following Section for charging procedure.

6   Take great care to protect the eyes and skin against accidental spillage of acid when holding the battery. Wear eyeshields at all times. Eyes contaminated with acid must be immediately flushed with fresh water and examined by a doctor. Similar attention should be given to a spillage on the skin.

## 8 Battery: charging

1   Battery life will be effectively shortened if the charge exceeds a rate of about 1.8 amp. Charge in a well ventilated area, check the side vent is clear and remove the cell caps, otherwise the gas created within the battery may burst the case with disastrous consequences. Do not charge the battery in place with its leads connected; this can damage the rectifier.

2   Check the charger connections are correct, red to positive (+), black negative (−). When refitting the battery, connect its black (negative) lead to earth otherwise the system will be permanently damaged.

3   Sulphuric acid is extremely corrosive. Wash hands promptly after handling the battery because its case is likely to be contaminated. Note the following precautions:

　　Do not allow smoking or naked flames near batteries.
　　Do avoid acid contact with skin, eyes and clothing.
　　Do keep battery electrolyte level maintained.
　　Do avoid over-high charge rates.
　　Do avoid leaving the battery discharged.
　　Do avoid freezing.
　　Do use only distilled or demineralised water for topping up.

## 9 Fuse: renewal

1   A plastic holder fitted to the battery positive lead contains the 20 amp fuse. Another holder fitted to the positive lead contains the spare fuse. Before replacing a blown fuse, check the system thoroughly to

trace and eliminate the fault. If no spare is available, a 'get you home' remedy is to remove the fuse, wrap it in silver paper and refit it. Never do this if there is evidence of an electrical fault otherwise more serious damage will result. Replace the 'doctored' fuse at the earliest opportunity and carry a spare.

6.1 A ballast resistor is fitted behind headlamp assembly on UK models

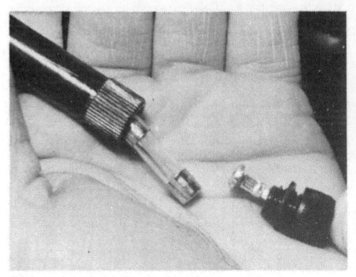

9.1 A plastic holder contains the 20 amp fuse

## 10 Bulbs: renewal

1   Apart from the headlamp sealed beam unit fitted to US models, all bulbs are of the bayonet type and can be released by pushing in, turning anti-clockwise and pulling from the holder. The tail/stop lamp is fitted with a double filament bulb which has offset pins to prevent unintentional reversal of its holder.

2   On UK models, gain access to the main headlamp bulb holder by detaching the rim, complete with reflector and glass, from the shell. Unclip the holder from the centre of the reflector to expose the bulb. Pull the pilot lamp holder from the reflector to expose the bulb.

3   On US models, detach the headlamp rim from the shell. Pull the rim forward and unplug the wiring connection from the rear of the sealed beam unit. Note the fitted positions of the unit, its retaining ring and the headlamp rim. Detach the ring from the rim and the sealed beam unit from the ring.

4   On all models, the tail/stop and direction indicator bulbs can be removed after detachment of the plastic lens, which will be secured by screws. Take care not to tear the lens seal; this keeps dirt and moisture away from the backplate and electrical contacts and must be renewed if damaged.

5   The instrument and warning light bulbs are fitted in rubber holders which can be unplugged from the base of the instrument or console once it is exposed.

6   Clean any corrosion or moisture from the holder and check its contacts are free to move when depressed. Fit the bulb and lens. A lens can be cracked if its securing screws are overtightened. After tightening the headlamp rim, check beam alignment.

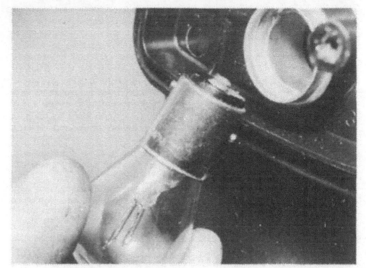

10.1 The tail/stop lamp bulb has offset pins to prevent reversal

10.2a On UK models, unclip the headlamp bulb holder ...

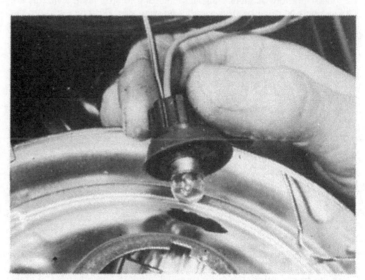

10.2b ... and pull the pilot lamp holder from the reflector

10.4 Check each lens seal is undamaged

10.5 Unplug the instrument bulb holders from the base of the console

## 11 Headlamp: beam alignment

1   UK regulations stipulate that the headlamp must be aligned so that the light will not dazzle a person standing at a distance greater than 25 feet from the lamp, whose eye level is not less than 3 feet 6 inches above that place. It is easy to approximate this setting by placing the machine 25 feet away from a wall, on a level road, and setting the dip beam height so that it is concentrated at the same height as the distance of the centre of the headlamp from the ground. The rider must be seated normally during this operation and also the pillion passenger, if one is carried regularly.

2   Most other areas have similar regulations controlling headlamp beam alignment. These should be checked before any adjustment is made.

3   To effect beam vertical alignment, detach the rim complete with reflector and glass, loosen the headlamp shell securing nuts and pivot the shell up or down as required.

## 12 Switches: testing

1   Before suspecting a switch, refer to Section 2 and check its circuit wiring.

2   The ignition, oil level, neutral indicator and stop lamp switches are all sealed and must be renewed if unserviceable. Dirty contacts are the cause of most troubles with handlebar switches; these can be cleaned with a special electrical contact cleaner. Internal breakage will necessitate switch renewal.

3 _. To test, refer to the accompanying wiring diagram and identify the switch and its circuit. Prevent the possibility of a short circuit by disconnecting the battery. Disconnect the switch, set a multimeter to its resistance function and check for continuity between the switch terminals (and earth, where necessary) whilst operating the switch. If no continuity is found in the positions shown for continuity in the diagram, the switch is defective.

4   It is dangerous to have a defective lighting switch whilst riding at night. Failure will plunge the rider into darkness with disastrous consequences.

## 13 Stop lamp switches: adjustment

1   The stop lamp is operated by two switches, front and rear. Only the rear switch is adjustable. If the lamp is late in operating, raise the switch body by turning its adjuster nut clockwise whilst holding the switch steady. If the lamp is permanently on, then lower its body. As a guide to operation, the lamp should illuminate immediately the brake pedal is depressed.

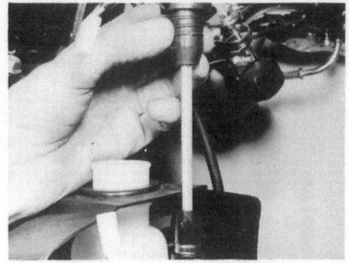

12.2a The oil level switch can be unplugged from the tank

12.2b Examine the handlebar switch contacts

13.1a The front stop lamp switch is plugged into the handlebar lever bracket

13.1b Rotate the adjuster nut to alter the rear stop lamp switch position

## 14 Flasher unit: renewal

1   If the unit is functioning correctly, a series of clicks will be heard when the indicator lamps are operating. If it malfunctions and the bulbs are serviceable, the usual symptom is one flash before the unit goes dead.

2   The unit is sealed; to renew it, unplug the electrical connectors and slide the unit from its mounting. Do not subject the new unit to sudden shock, it is easily damaged. Fitting is a direct reversal of removal.

## 15 Horn: adjustment and renewal

1   Volume adjustment is provided by means of a screw at the rear of the horn case. Turn the screw fractionally to increase volume.

2   The unit is sealed; to renew it, disconnect the electrical wires from its terminals, having noted their fitted positions, and unbolt the unit from its mounting. Fitting is a direct reversal of removal.

## 16 Oil level warning light circuit: testing

1   The oil level warning light set in the instrument panel is operated by a float-type switch mounted in the oil tank and gives warning of a dangerously low engine oil level. It is connected to the neutral indicator light circuit via a diode contained in the switch so that the oil level lamp lights (as a check of the bulb's condition) whenever the transmission is in neutral. As soon as the machine is put in gear both lights should go out and the oil level light should not come on unless there is 200 cc of oil or less left in the tank.

2   If the light fails to work first check the bulb itself. If the bulb is functioning correctly remove the fuel tank and disconnect the switch three-pin connector; the neutral indicator light should function normally.

3   To check the wiring, bridge the Black/red and Black/yellow wire terminals on the main harness half of the switch connector plug. With

14.1 The flasher unit is located in the battery compartment

the ignition switched on, the lamp will light; if not check that there is full battery voltage at the bulb, then at the connector Black/red wire terminal. If the supply is proved good, check the earth return from the Black/yellow wire terminal, using the methods outlined in Sections 2 and 3 of this Chapter.

4   If the wiring and bulb are sound and the fault persists, the switch must be faulty and should be renewed; repairs are not possible. The diode on the neutral indicator circuit connection can be checked by measuring the resistance across the Light green and Black/red wire terminals, first in one direction, then in the other. Continuity should only be found in one direction; if it is found in both directions, or in neither, then the diode is faulty and must be renewed as part of the sealed switch unit.

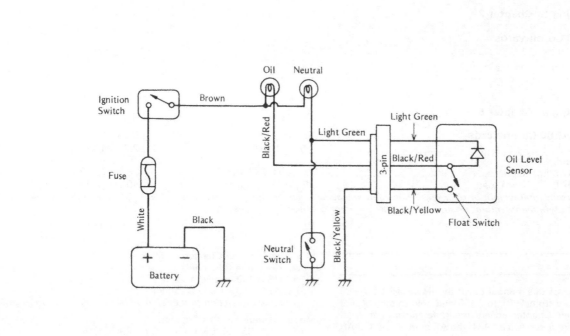

Fig. 6.2 Oil level warning light circuit diagram

# Chapter 7 The 1984 on models

## Contents

## Specifications

*The following specifications apply to the UK AR50 and 80 models produced since 1984, namely the C2, C3, C4, C5, C6, C7, C8, C9 and C10 models. Information is given only where different from that shown for the AR50 and 80 (C1) models in the Specifications Sections of Chapters 1 to 6. Unless otherwise specified, information applies to all later models. Where information is not available, the abbreviation N/Av will be found.*

### Specifications relating to Chapter 1

#### Engine
Compression pressure @ 5 kicks, engine fully warmed up and throttle fully open:

| | |
|---|---|
| AR50 ........................................................... | 7.5 – 11.7 kg/cm² (107 – 166 psi) |
| AR80 ........................................................... | 8.8 – 13.5 kg/cm² (125 – 192 psi) |

#### Piston
Piston to bore clearance:

| | |
|---|---|
| AR50 ........................................................... | 0.030 – 0.040 mm (0.0012 – 0.0016 in) |
| AR80 ........................................................... | 0.035 – 0.045 mm (0.0014 – 0.0018 in) |

#### Gearbox

| | |
|---|---|
| Gearchange mechanism spring free length ............... | N/Av |
| Secondary reduction ratio – AR80 C7 and C8 ............ | 2.867:1 (43/15T) |

| Torque wrench settings | kgf m | lbf ft |
|---|---|---|
| Clutch hub bolt – AR50/80 C6 (apply oil to bolt threads and seating) ............................................ | 2.8 | 20.0 |
| Clutch spring bolts – AR50/80 C6 onwards ............... | 0.4 | 2.9 |
| Gearchange set lever pivot screw – AR50/80 C7 onwards (apply locking agent to threads) ............... | 0.45 | 3.3 |
| Gearchange drum camplate bolt – AR50/80 C7 onwards (apply locking agent to threads) ............... | 1.2 | 8.7 |

### Specifications relating to Chapter 2

#### Carburettor – AR80 C6 onwards

| | |
|---|---|
| Needle jet .................................................... | O – 4 |
| Jet needle .................................................... | 4CL15 – 3 |
| Pilot air screw .............................................. | 1¼ turns out from fully closed |

### Specifications relating to Chapter 5

#### Tyres – AR50 and AR80 C5 onwards

| | Front | Rear |
|---|---|---|
| Size ............................................................ | 2.50 – 18 4PR | 2.75 – 18 6PR |
| Pressures – tyres cold: | | |
| Up to 215 lb (97.5 kg) load ........................... | 25 psi (1.75 kg/cm²) | 28 psi (2.00 kg/cm²) |
| 215 – 384 lb (97.5 – 174 kg) load ................. | 25 psi (1.75 kg/cm²) | 40 psi (2.80 kg/cm²) |

*Loads quoted represent total weight of rider, passenger, and any accessories or luggage. Pressures apply to OE tyres only – if different tyres are fitted check with tyre manufacturer or supplier whether different pressures are necessary.*

## 1  Introduction

The first six Chapters of this manual cover the Kawasaki AE/AR50 and 80 models *produced* from 1981 to 1983 and sold in the UK and US. This supplementary Chapter covers the differences in the UK AR50 and 80 models *produced* since 1984; the AE models and US AR models remained unchanged until they were discontinued.

To avoid confusion over model coverage, all models are listed below, grouped according to their coverage in this manual.

Models covered in Chapter 1 to 6:
All AE50 and 80 models
All US AR50 and 80 models
UK AR50/80 A1 and C1 models
*Information relating to model identification is given on page 6*

Models covered in this Chapter:
AR50 C2, C3, C4, C5, C6, C7, C8, C9 and C10
AR80 C2, C3, C4, C5, C6, C7 and C8

When working on one of these later models, refer first to this

Chapter; if there are no changes in relevant specifications or working procedures, then the task is substantially the same as that described in the relevant part of Chapters 1 to 6.

None of the later models received any major alteration, even styling changes being restricted to the usual alterations to paintwork and graphics. In many cases, components have been changed but are not mentioned because working procedures or specifications remain the same; do not assume that this means that such components are interchangeable. Always take great care to identify your machine exactly when ordering replacement parts. Identification details and the initial frame and engine numbers with which each machine's production run commenced are given below.

**AR50 C2** – Production year 1984, imported into the UK from November 1983 to February 1985. Frame number AR050C-001001 on. Engine number AR050AE035801 on. Available in Ebony or Lime Green, this model is virtually identical to the C1 described in Chapters 1 to 6.

**AR50 C3** – Production year 1985, imported into the UK from October 1984 to January 1986. Frame number AR050C-002501 on. Engine number AR050AE040501 on. Available in Ebony or Lime Green.

**AR50 C4** – Production year 1986, imported into the UK from January 1986 to March 1987. Frame number AR050C-004101 on. Engine number AR050AE045001 on. Available in Ebony or Lime Green.

**AR50 C5** – Production year 1987, imported into the UK from March 1987 to January 1988. Frame number AR050C-005301 on. Engine number AR050AE048501 on. Available in Ebony or Polar White.

**AR50 C6** – Production year 1988/9, imported into the UK from January 1988 to February 1989. Frame number AR050C-007001 on. Engine number AR050AE050501 on. Available in Ebony or Polar White.

**AR50 C7** – Production year 1989, imported into the UK from February 1989 to October 1990. Frame number AR050C-009201 on. Engine number AR050AE054201 on. Available in Ebony or Polar White.

**AR50 C8** – Production year 1990, imported into the UK from April 1990 to October 1991. Frame number AR050C-013001 on. Engine number AR050AE059001 on. Available in Ebony or Polar White.

**AR50 C9** – Production year 1991/92, imported into the UK from April 1991 to November 1992. Frame number AR050C-018701 on. Engine number details are not available. Available in Ebony or Polar White.

**AR50 C10** – Production year 1993-95, imported into the UK from November 1992. Frame number AR050C-024001 on. Engine number details are not available. Ebony or Sunbeam Red colours.

**AR80 C2** – Production year 1984, imported into the UK from January 1984 to August 1985. Frame number AR080C-000701 on. Engine number AR080AE029501 on. Available in Sunbeam Red or Ebony, this model is virtually identical to the C1 described in Chapters 1 to 6.

**AR80 C3** – Production year 1985, imported into the UK from October 1984 to March 1987. Frame number AR080C-001701 on. Engine number AR080AE032001 on. Available in Sunbeam Red or Ebony.

**AR80 C4** – Production year 1986, imported into the UK from August 1986 to March 1987. Frame number AR080C-002401 on. Engine number AR080AE033301 on. Available in Ebony or Polar White.

**AR80 C5** – Production year 1987, imported into the UK from March 1987 to January 1988. Frame number AR080C-002801 on. Engine number AR080AE034501 on. Available in Ebony or Polar White.

**AR80 C6** – Production year 1988, imported into the UK from January 1988 to January 1989. Frame number AR080C-003501 on. Engine number AR080AE035501 on. Available in Ebony or Polar White.

**AR80 C7** – Production year 1989, imported into the UK from January 1989 to March 1990. Frame number AR080C-004601 on. Engine number AR080AE042001 on. Available in Ebony or Polar White.

**AR80 C8** – Production year 1990, imported into the UK from April 1990 to November 1992 when discontinued. Frame number AR080C-006001 on. Engine number AR080AE046001 on. Available in Ebony or Polar White.

## 2 Routine maintenance: general

**Prop stand retracting mechanism**

1 All later models are fitted with an automatic prop stand retracting

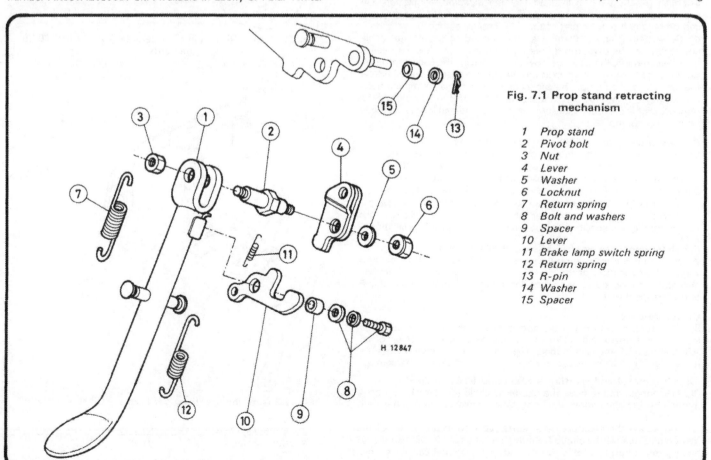

**Fig. 7.1 Prop stand retracting mechanism**

1  Prop stand
2  Pivot bolt
3  Nut
4  Lever
5  Washer
6  Locknut
7  Return spring
8  Bolt and washers
9  Spacer
10  Lever
11  Brake lamp switch spring
12  Return spring
13  R-pin
14  Washer
15  Spacer

H 12847

mechanism which is operated by a short cable branching off the main clutch cable. Whenever the clutch is adjusted, check the adjustment of this mechanism, which should be set so that all cable free play is just eliminated when the stand is in the down position. Secure the cable adjuster locknut and apply a few drops of oil to the mechanism pivots, cable nipples and bearing surfaces.

### Oil pump synchronisation – all models

2 The oil pump cable setting **must** be checked, and reset if necessary, whenever the throttle cable is disturbed, as well as at the specified interval.

3 The instructions given in Routine maintenance describe oil pump synchronisation as being carried out at idle on UK models and at full throttle for US models. While this remains applicable for all later models, owners should note that in practice the oil pump pulleys all appear to have both marks inscribed. In this case, the synchronisation can be double-checked at both throttle openings (switch off the engine to make the full throttle check!). *Note that if there is any discrepancy, it is preferable to use the full throttle mark as this more closely approximates normal running conditions.*

### 3  Engine: modifications

### Piston rings – AR80 C6, C7 and C8

1 Note that this model is fitted with a thin expander underneath the second (lower) piston ring. It is removed and refitted in the same way as the rings themselves, but is considerably more flexible. Clean it carefully and check for signs of cracking, fatigue or damage; it can be renewed only as part of the piston ring set. On reassembly, fit the expander ring ends on each side of the ring locating peg.

2 In addition, the top piston ring is of the Keystone type, being slightly tapered in section. If the rings are removed, always ensure that this ring is fitted to the top groove of the piston.

### Tachometer drive mechanism

3 If the tachometer drive mechanism is found to be damaged or worn, seek the advice of a good authorised Kawasaki dealer to establish exactly what is available for the machine in question. On some models the mechanism is available separately, but on others it can be obtained only as part of the oil pump assembly.

4 On AR50 C9 and C10 models, the method of tachometer cable connection at the oil pump differs from previous models. To release the cable, remove the oil pump cover and displace the cable grommet. Remove the two screws which clamp the cable in position and withdraw the cable from the pump. The cable connection at the instrument head is unchanged from previous models. Refit the cable in a reverse of the removal procedure, but make sure that its squared end locates correctly with the oil pump.

### Primary drive gear – AR50 and 80 C6 onwards

5 On these models the primary drive gear shock absorber is deleted. The drive pinion is now located by a Woodruff key and secured by a locking nut and washer. Comparison of the accompanying illustration with Fig. 1.3 will show the layout of the modified assembly.

6 Note that the threads of the crankshaft must be lightly oiled before refitting the nut. Prevent the crankshaft from turning whilst the nut is tightened to the specified torque wrench setting (see Specifications).

### Flywheel generator rotor – AR50 and 80 C7 onwards

7 Note that the rotor is now located on the crankshaft taper by a Woodruff key instead of the dowel pin used previously. The removal and refitting procedures remain as described in Chapter 1, Sections 9 and 29 respectively.

### Clutch release

8 With reference to Fig. 1.14, note that the clutch release lever is located by a dowel pin set in the crankcase right-hand cover of all AR50 and 80 C2 and later models. The pin must be removed before the release lever can be extracted, and must not be omitted on reassembly.

### Clutch removal and refitting – AR50 and 80 C7 onwards

9 Disconnect and remove the operating cable (Chapter 1, Section 4, paragraph 11) and remove the right-hand crankcase cover (Chapter 1, Section 7).

10 Withdraw the steel ball and centre piece from the centre of the clutch and remove the circlip from the end of its shaft. Slacken the six spring retaining plate bolts evenly and in a diagonal sequence, then remove them complete with the plate and springs. The end plate,

friction and plain plates and the clutch hub can then be drawn off the shaft as an assembly. The clutch drum (with integral kickstart driven pinion) is positioned on the shaft by two circlips; remove the first circlip, followed by the thrust washer, and withdraw the drum. A further thrust washer and the second circlip are located behind the drum. Note that the latter must be removed if input shaft removal is contemplated.

11 Refer to Chapter 1, Section 22 for details of examination and renovation.

12 Refitting is a straightforward reversal of the removal sequence, noting the following points: If any of the circlips show signs of fatigue or damage renew them. When installing the six spring plate retaining bolts tighten them evenly and in a diagonal sequence.

### Kickstart driven pinion

13 With reference to Fig. 1.14, note that the kickstart driven pinion (item 2) is shown as having two dogs which engage with slots in the back of the clutch outer drum central boss. On AR50 and 80 C2, C3, C4 and C5 models the number of dogs and slots was increased to four for greater strength. This meant the fitting of a modified driven pinion and clutch outer drum; note that these can be fitted only as a matched pair to earlier models.

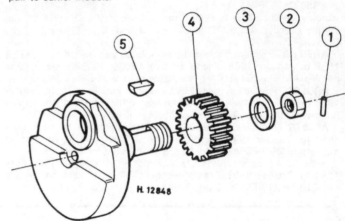

**Fig. 7.2 Primary drive pinion assembly – AR50 and 80 C6 onwards**

1  Oil pump drive pin       4  Drive pinion
2  Nut                      5  Woodruff key
3  Washer

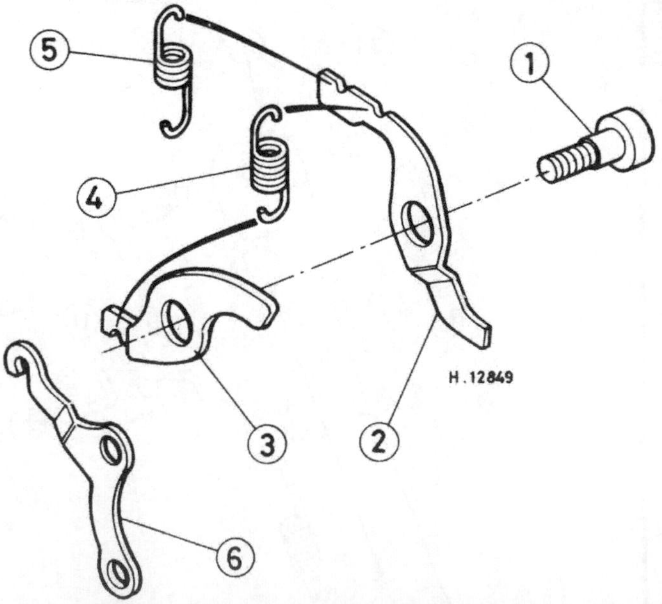

**Fig. 7.3 Gearchange drum set levers – C2 to C6 models**

1  Screw                 5  Spring*
2  Neutral set lever      6  Extended bearing retainer
3  Gear set lever            plate*
4  Spring

*\* not fitted to C6 models*

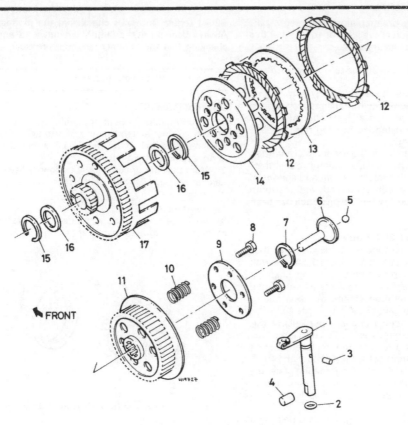

**Fig. 7.4 Clutch – AR50 and 80 C7 onwards**

| | | | |
|---|---|---|---|
| 1 | Release lever | 6 | Centre piece | 10 | Spring – 6 off | 14 | Clutch hub |
| 2 | O-ring | 7 | Circlip | 11 | End plate | 15 | Circlip – 2 off |
| 3 | Dowel pin | 8 | Bolt – 6 off | 12 | Friction plate – 4 off | 16 | Thrust washer – 2 off |
| 4 | Pushrod | 9 | Spring retaining plate | 13 | Plain plate – 3 off | 17 | Clutch drum |
| 5 | Steel ball | | | | | | |

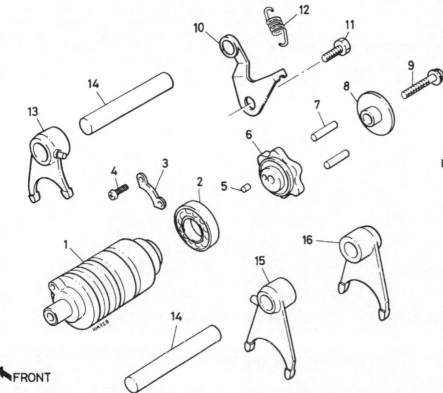

**Fig. 7.5 Gearchange drum assembly –
AR50 and 80 C7 onwards**

1  Gearchange drum
2  Bearing
3  Retaining plate
4  Screw – 2 off
5  Dowel pin
6  Camplate
7  Change pin – 6 off
8  Pin retaining plate
9  Allen bolt
10  Set lever
11  Pivot bolt
12  Spring
13  Selector fork
14  Selector fork shaft – 2 off
15  Selector fork
16  Selector fork

14  On AR50 and 80 C6 on models, the driven pinion is an integral part of the clutch outer drum. No information is available as to whether this single component can be fitted to earlier models.

### Gearchange mechanism external components – AR50 and 80 C2 to C6 models

15  The gearchange drum set levers have received two further modifications; comparison of the accompanying illustration with Fig. 1.13 will show the basic difference in layout. Note that AR50 and 80 C6 models are not fitted with the extended bearing retainer plate and additional spring of the earlier models.

16  Before disturbing the assembly examine it closely and make written notes of how each lever is arranged (particularly which is outermost) and of how each lever engages with the selector drum and spring(s). Be careful to ensure that the levers are refitted correctly and that both are free to move (against spring pressure) when the pivot bolt has been securely tightened.

### Gearchange mechanism – AR50 and 80 C7 onwards

17  The gearchange drum set lever arrangement has been further modified on these models. A single set lever is fitted which locates with the selector drum camplate and is held in tension by a spring, linking it to an extended input shaft bearing retaining plate. Note its exact fitted position prior to removal and when refitting ensure that it is free to move against spring pressure when its pivot bolt is tightened.

18  The gearchange drum has also been modified, particularly the camplate components on its right-hand end. To dismantle the camplate assembly, first disengage and remove the gearchange shaft arm from the drum and remove the drum set lever. Remove the Allen bolt from the end of the drum and withdraw the pin retaining plate, the six change pins and the camplate. Note that the camplate is located on the drum end by a small dowel pin; ensure that this is correctly positioned on reassembly.

19  The drum right-hand end resides in a bearing, which is held by a retaining plate on the inside of the crankcase. If separating the crankcases for drum removal note that the camplate components must be removed before the drum can be withdrawn from the right-hand crankcase half.

### Neutral indicator switch – AR50 and 80 C6 onwards

20  On these models note that a number of 0.5 mm (0.020 in) thick shims may be fitted under the neutral indicator switch sealing washer, to adjust the distance of the switch from the selector drum. If the switch is ever disturbed make a careful note of the position and number of shims and be careful to refit them exactly as they were found.

### Gearbox breather – AR50 and 80 C6 onwards

21  On later models the gearbox breather was moved from the top rear corner of the crankcase mating surface to the top surface of the crankcase left-hand half, underneath the cylinder barrel. A plastic union and 360 mm (14 in) long length of hose are fitted to channel away any oily residue which might be emitted.

22  At regular intervals check that the pipe is clean and unblocked. Always clean it out thoroughly whenever the engine unit is dismantled, checking that the breather passages are clear.

### 4  Instrument drives: general

**Speedometer drive**

1  Later models have a modified speedometer drive assembly as shown in the accompanying illustration. Removal of the oil seal remains as described in Chapter 5.6, although note that the drive gear is now retained by a circlip.

2  Note that the driven gear is no longer listed as a separate part and can only be obtained with the complete speedometer gearbox assembly.

**Tachometer drive**

3  Note the modifications made to the tachometer drive on later models in Section 3, paragraphs 3 and 4 of this Chapter.

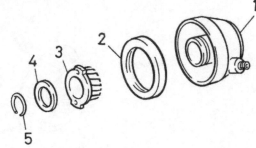

**Fig. 7.6 Speedometer drive gear assembly**

| | | | |
|---|---|---|---|
| 1 | Housing and driven gear | 4 | Washer |
| 2 | Oil seal | 5 | Circlip |
| 3 | Drive gear | | |

### 5  Instrument panel bulbs: renewal

1  Later models may be fitted with instrument panel bulbs of the capless type, which are a simple press fit in their bulb holders. On renewing bulbs in the instrument panel, first detach the bulb holder as described in Chapter 6, then check to see whether the bulb has a metal base and is of the bayonet fitting (ie press in and twist anticlockwise to release) or whether it is of the capless type, with a plain glass envelope, which can be easily pulled out of its holder. **Do not** use force or the bulb may break, with the consequent risk of personal injury.

2  To refit capless bulbs, align their wire tails with the metal contacts in the bulb holder and press them into place. Take care not to damage the fine wire tails.

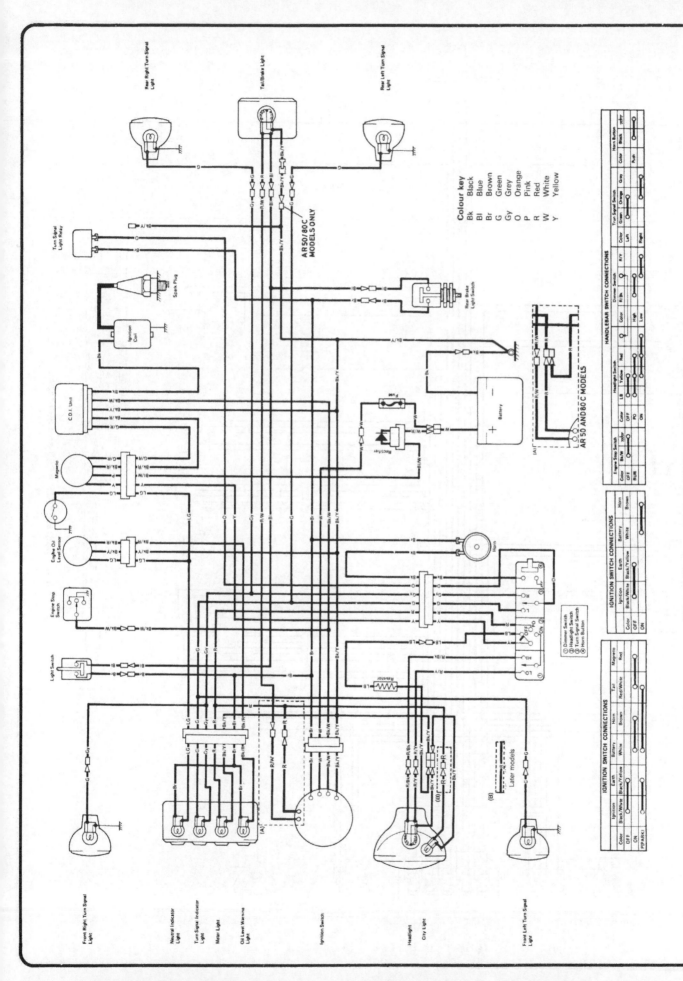

Wiring diagram – UK AE/AR50 and 80 models

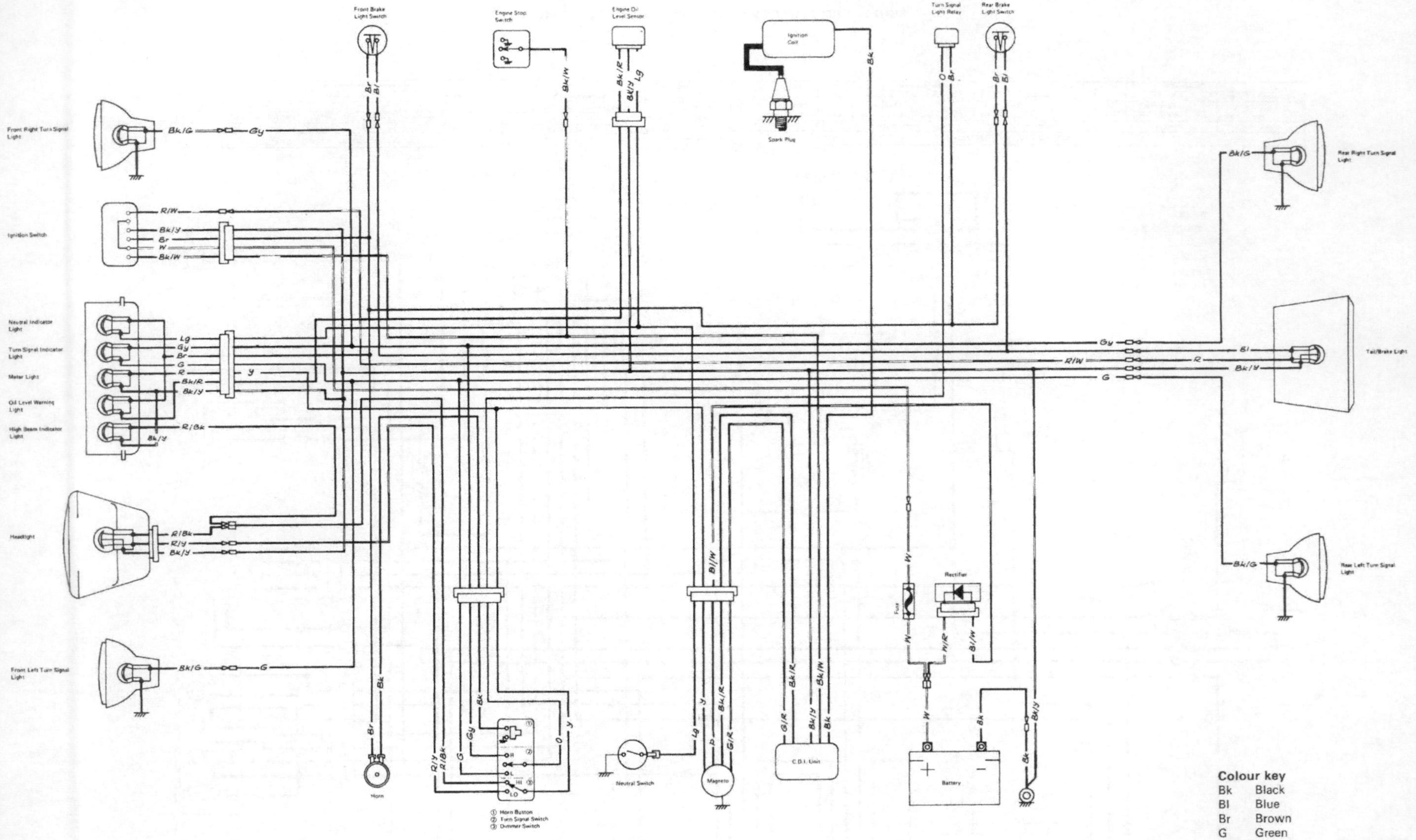

Rear Right Turn Signal Light

Tail/Brake Light

Rear Left Turn Signal Light

**Colour key**

| | |
|---|---|
| Bk | Black |
| Bl | Blue |
| Br | Brown |
| G | Green |
| Gy | Grey |
| O | Orange |
| P | Pink |
| R | Red |
| W | White |
| Y | Yellow |

Rear Brake Light Switch

Turn Signal Light Relay

Rectifier

Battery

Ignition Coil

Spark Plug

Fuse

C.D.I. Unit

Magneto

Engine Oil Level Sensor

Neutral Switch

Engine Stand Switch

Horn Button
Turn Signal Switch
Dimmer Switch

Front Brake Light Switch

Horn

**HANDLEBAR SWITCH CONNECTIONS**

Engine Stop Switch

| Color | Black | White |
|---|---|---|
| OFF | o | |
| Run | o | o |

Horn Button

| Color | Black | Brown |
|---|---|---|
| Push | o | o |

Turn Signal Switch

| Color | Green | Orange | Grey |
|---|---|---|---|
| Left | o | o | |
| Right | | o | o |

Dimmer Switch

| Color | Red/Yellow | Red/Black |
|---|---|---|
| High | o | |
| Low | | o |

**IGNITION SWITCH CONNECTIONS**

| | Tail | | |
|---|---|---|---|
| | Brown | Red/White | |

| Ignition | Black/Yellow | Black/White | Red/Black |
|---|---|---|---|
| OFF | | | |
| ON | | | |
| PARK | | | |

**Wiring diagram – US AR50 and 80 models**

Front Right Turn Signal Light

Ignition Switch

Neutral Indicator Light

Turn Signal Indicator Light

Meter Light

Oil Level Warning Light

High Beam Indicator Light

Headlight

Front Left Turn Signal Light

# English/American terminology

Because this book has been written in England, British English component names, phrases and spellings have been used throughout. American English usage is quite often different and whereas normally no confusion should occur, a list of equivalent terminology is given below.

| English | American | English | American |
| --- | --- | --- | --- |
| Air filter | Air cleaner | Number plate | License plate |
| Alignment (headlamp) | Aim | Output or layshaft | Countershaft |
| Allen screw/key | Socket screw/wrench | Panniers | Side cases |
| Anticlockwise | Counterclockwise | Paraffin | Kerosene |
| Bottom/top gear | Low/high gear | Petrol | Gasoline |
| Bottom/top yoke | Bottom/top triple clamp | Petrol/fuel tank | Gas tank |
| Bush | Bushing | Pinking | Pinging |
| Carburettor | Carburetor | Rear suspension unit | Rear shock absorber |
| Catch | Latch | Rocker cover | Valve cover |
| Circlip | Snap ring | Selector | Shifter |
| Clutch drum | Clutch housing | Self-locking pliers | Vise-grips |
| Dip switch | Dimmer switch | Side or parking lamp | Parking or auxiliary light |
| Disulphide | Disulfide | Side or prop stand | Kick stand |
| Dynamo | DC generator | Silencer | Muffler |
| Earth | Ground | Spanner | Wrench |
| End float | End play | Split pin | Cotter pin |
| Engineer's blue | Machinist's dye | Stanchion | Tube |
| Exhaust pipe | Header | Sulphuric | Sulfuric |
| Fault diagnosis | Trouble shooting | Sump | Oil pan |
| Float chamber | Float bowl | Swinging arm | Swingarm |
| Footrest | Footpeg | Tab washer | Lock washer |
| Fuel/petrol tap | Petcock | Top box | Trunk |
| Gaiter | Boot | Torch | Flashlight |
| Gearbox | Transmission | Two/four stroke | Two/four cycle |
| Gearchange | Shift | Tyre | Tire |
| Gudgeon pin | Wrist/piston pin | Valve collar | Valve retainer |
| Indicator | Turn signal | Valve collets | Valve cotters |
| Inlet | Intake | Vice | Vise |
| Input shaft or mainshaft | Mainshaft | Wheel spindle | Axle |
| Kickstart | Kickstarter | White spirit | Stoddard solvent |
| Lower leg | Slider | Windscreen | Windshield |
| Mudguard | Fender | | |

# Index